Engineering Design

Prentice Hall Modular Series for Engineering

Now, you can tailor your course materials to satisfy the needs of you and your students. With the Prentice Hall Modular Series for Engineering, you can mix and match these concise, inexpensive books for your classes. They are ideal for courses in which a variety of languages and software are being covered. All books focus on using real-world applications to help motivate your course.

Current Modules Include:

Introduction to MATLAB for Engineers and Scientists
    Delores Etter – University of Colorado, Boulder
    1996, 145 pp. 0-13-519703-1

Introduction to C++ for Engineers and Scientists
    Delores Etter – University of Colorado, Boulder
    1997, 160 pp. 0-13-254731-7

Introduction to ANSI C for Engineers and Scientists
    Delores Etter – University of Colorado, Boulder
    1996, 164 pp. 0-13-241381-7

Introduction to Fortran 90 for Engineers and Scientists
    Larry Nyhoff and Sanford Leestma – both of Calvin College
    1997, 336 pp. 0-13-505215-7

Fundamentals of AutoCAD
    Mark Dix and Paul Riley – CAD Support Associates
    1998, 250 pp. 0-13-860362-6

Engineering Design: A Day in the Life of Four Engineers
    Mark Horenstein – Boston University
    1998, 150 pp. 0-13-8660242-8

Introduction to the Internet for Engineers and Scientists
    Scott James – GMI Institute
    1998, 150 pp. 0-13-856691-7

Introduction to ProENGINEER
    Jeffrey Freeman and Andrew Whelan – both of the University of Iowa
    1998, 150 pp. 0-13-861048-7

Future Modules Include:

Engineering Ethics
Engineering Problem-Solving
Introduction to MS Word
Introduction to WordPerfect
Introduction to Excel
Introduction to Lotus 1-2-3

Engineering Design: A Day in the Life of Four Engineers

MARK N. HORENSTEIN

Boston University

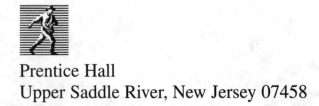

Prentice Hall
Upper Saddle River, New Jersey 07458

Library of Congress Cataloging-in-Publication Data

Horenstein, Mark N.
    Engineering design : a day in the life of four engineers / Mark N. Horenstein.
      p.   cm. — (Prentice Hall modular series for engineering)
    Includes index.
    ISBN 0-13-660242-8
    1. Engineering design. 2. Automobiles, Electric—Design.
I. Title. II. Series.
TA174.H67 1998
620'.0042—dc21

                        97-17165
                        CIP

Editor-in-chief: **Marcia Horton**
Acquisitions editor: **Eric Svendsen**
Director of production and manufacturing: **David W. Riccardi**
Managing editor: **Bayani Mendoza DeLeon**
Production editor: **Judy Winthrop**
Cover director: **Jayne Conte**
Manufacturing buyer: **Julia Meehan**
Editorial assistant: **Andreau Au**

Copyright © 1998 by Prentice-Hall, Inc.
Simon & Schuster/A Viacom Company
Upper Saddle River, New Jersey 07458

The author and publisher of this book have used their best efforts in preparing this book. These efforts include the development, research, and testing of the theories and programs to determine their effectiveness. The author and publisher shall not be liable in any event for incidental or consequential damages in connection with, or arising out of, the furnishing, performance, or use of these programs.

Printed in the United States of America

10 9 8 7 6 5 4 3 2 1

ISBN 0-13-660242-8

PRENTICE-HALL INTERNATIONAL (UK) LIMITED, *London*
PRENTICE-HALL OF AUSTRALIA PTY. LIMITED, *Sydney*
PRENTICE-HALL CANADA, INC., *Toronto*
PRENTICE-HALL HISPANOAMERICANA, S.A., *Mexico*
PRENTICE-HALL OF INDIA PRIVATE LIMITED, *New Delhi*
PRENTICE-HALL OF JAPAN, INC., *Tokyo*
SIMON & SCHUSTER ASIA PTE., LTD., *Singapore*
EDITORA PRENTICE-HALL DO BRAZIL, LTDA., *Rio de Janeiro*

Contents

Preface

Engineering Design has been written for engineering undergraduates at all levels, from first year through senior year students. Each chapter describes a day or event in the lives of four engineers and also contains additional examples, exercises, and open-ended design problems for the student of engineering design. Design problems may be formulated on paper or, where appropriate, constructed from available materials. The text is easily integrated into design-oriented introductory or capstone courses at all levels. The topics chosen for coverage represent a sampling of the principles and foundations of engineering design. These topics are intended to supplement other texts and resource materials that provide a more detailed understanding of basic fundamentals.

Several individuals contributed to the ideas and concepts presented in *Engineering Design*. Professors Thomas Bifano and James Bethune provided guidance with aspects of the book related to mechanical engineering, and Prof. Michael Ruane, my colleague in the teaching of design at Boston University, contributed to many of the concepts relating to logbook practices, project execution, and technical writing. Special thanks to Professor Bethune for his help with Autocad drawings. Several of the concepts presented in this book were offshoots of a workshop on engineering design held bi-annually by Professor Charles Lovas of Southern Methodist University. Martin Lurie provided much of the inspiration for the organizational aspects of Xebec Research and Development.

ENGINEERING DESIGN: A DAY IN THE LIFE OF FOUR ENGINEERS

Synopsis:

Imagine yourself several years from now, after you've finished your college studies in engineering. What will life be like on your new job? How will everything you learn in school relate to your work and your career? What do engineers really do? This book will provide you with a vision of the future while helping to teach you the important principles of engineering design. Told from the point of view of four different types of engineers working together on a design team, the book describes the problems they encounter and the tools used to solve them while working on an interdisciplinary engineering design project. Each chapter is filled with additional examples, problems, and open-ended exercises for students at all levels of engineering design.

The four engineers:

Sally-Electrical Engineer
Dave-Mechanical Engineer
Anne-Computer Engineer
Keith-Industrial Engineer

The company:

Xebec Research and Development (pronounced Zee-beck)—Hired by a team of investors to design a prototype for a mass-produced electric automobile.

Contents:

**CHAPTER 1 SO YOU'VE FOUND THE CAFETERIA: FIRST DAY ON
THE JOB**

This chapter introduces the engineers, describes the company and its various divisions, and explains the functions of each engineer within the design team. The context is an orientation session run for new employees.

**CHAPTER 2 GENERATING SOLUTIONS TO ENGINEERING
PROBLEMS: THE ART OF BRAINSTORMING**

The Engineering Manager describes the design assignment, then explains and leads the team through a brainstorming session. He describes the creative mind and explains how brainstorming generates new ideas and innovative approaches to problems. He also discusses the importance of well known and tested technologies as a starting point for good engineering design. By the end of the session, several new ideas applicable to the electric automobile emerge.

**CHAPTER 3 LOGBOOKS, SKETCHES, AND COMPUTER AIDED
DESIGN**

This chapter explains the use and content of engineering logbooks as used in real engineering companies. Dave, the mechanical engineer, records some new ideas about the front suspension system for the electric car. He prepares hand sketches of several possible designs, then teams up with Keith to produce electronic versions of the drawings using the company's computer-aided design software. He discusses with Keith how the electronic files can be integrated into an automated manufacturing process. Keith later does a fabrication/assembly cost analysis on the finished part.

CHAPTER 4 THE IMPORTANCE OF ESTIMATION AND HAND CALCULATIONS

Sally, the electrical engineer, designs a circuit for the electric power controller for the drive motors of the new vehicle. She analyzes the power requirements on paper, making a first cut at the circuit design using rough hand calculations, some basic physics, and her understanding of electric power circuits. Later that evening, the four junior engineers get into a discussion of how much paint will be required to manufacture the cars. The discussion again highlights the importance of estimation, analysis, and a feeling for real numbers in engineering design.

CHAPTER 5 COMPUTER USE IN ENGINEERING DESIGN

Anne writes software for the control panel of the new vehicle. She first discusses with Sally the merits of writing the program in assembly code, C++, FORTRAN, or MAT-LAB, and decides upon the best alternative before proceeding. The two then review Anne's flowchart for the control program and talk about the data communication links between Anne's control panel and Sally's motor controller.

CHAPTER 6 MECHANICAL LOADING AND TESTING

Dave shows the team the results of loading and stress tests on a set of materials proposed for possible use in constructing the chassis frame. He describes the tests, both destructive and non-destructive, then discusses how these tests might help him choose the proper material for best strength and durability at the lowest possible manufacturing cost. The chapter also highlights proper data organization and oral presentation skills.

CHAPTER 7 BREADBOARDING AND TESTING

Sally models her electric power controller on the company's simulation software and fine tunes component values. She builds and tests the first version of the circuit on a component breadboarding system, noting a few anomalies that the simulator did not identify. After several design iterations, she wire-wraps a working prototype and runs it through a series of performance tests.

CHAPTER 8 THE ROLE OF FAILURE IN ENGINEERING DESIGN

The team reviews the test results of the first prototype trial, learning how to improve their vehicle by observing what went wrong with the first version. Harry, the engineering manager, engages the discouraged junior engineers in a discussion of the classic engineering failures throughout history, focusing on the role of failure in engineering design. The design team resolves to revisit their design activities with renewed fervor.

CHAPTER 9 LEARNING TO WRITE AS AN ENGINEER

The four engineers attend a company-sponsored workshop on writing and communication skills. The workshop includes hands-on practice in writing memos, progress reports, proposals, and technical documentation.

ENGINEERING DESIGN: A DAY IN THE LIFE OF FOUR ENGINEERS

PROLOGUE

Imagine yourself several years from now, after you've finished your college studies in engineering. What will life as an engineer be like? How will everything you learn in school relate to your work and your career? What do engineers really do? This book will provide you with a vision of the future while helping to teach you the important principles of engineering design. Told from the point of view of four different types of engineers working together on a design team, the book describes the problems and tasks they encounter while working on an interdisciplinary engineering design project.

The activities center around a fictitious company, Xebec Research and Development Corporation (pronounced Zee-beck). Xebec has been hired by a team of investors to design the prototype for a mass-produced electric automobile. The story centers around the experiences and activities of four engineers: Sally, an electrical engineer specializing in electronic circuit design; Dave, a mechanical engineer familiar with materials, loading, and testing; Anne, a computer engineer with a background in software, digital hardware, and microprocessors; and Keith, an industrial engineer hired to develop the production process that will eventually be used to manufacture the automobile. These four engineers engage daily in the principal job of the engineer: design.

WHAT IS DESIGN?

Design involves the creation of a device, component, mechanical system, electronic circuit, software program, or engineering process that meets a set of desired specifications. The design engineer makes choices based on a thorough understanding of engineering fundamentals, feasibility constraints, reliability, cost, manufacturability, ergonomics, and human factors. Good design requires experience, knowledge, and a considerable amount of intuition. Although design knowledge can be learned from books, design experience and intuition are more readily acquired through practice, practice, and more practice. Repetition, apprenticeship, and hard work are the keys to learning design. Developing good design skills also requires "seasoning"—the process by which a novice engineer gradually learns the "rules of thumb" and "tricks of the trade" from other, more experienced engineers. Such design lore is usually conveyed by a tradition of methods and procedures that is passed orally, from one generation of engineers to the next. The experience of failure and subsequent revision is also a very important part of the design process. When the first attempt at a design fails, the engineer gains valuable insight into what changes and alterations may be needed to make the design successful. Thorough design involves testing prototypes, studying failures, and observing the results of design decisions. Only by practicing real design in a real engineering environment can an engineer truly learn design.

Design occurs whenever the human brain combines its knowledge of facts and figures with its creative abilities to synthesize a device, system, or procedure. Wisdom and intuition become important ingredients in this process. As an engineering student, you acquire the knowledge component of design via a carefully selected set of courses, both technical and non-technical. The creative component of design comes from allowing yourself to think about problems and their likely solutions. The part of design involving wisdom and intuition is acquired though time and devotion to the engineering profession.

The purpose of this book is to help you learn the principles of design. Written from the point of view of four junior engineers, it will allow you to vicariously experience life in a real engineering company and become familiar with common engineering practice. Each chapter includes exercises and open-ended problems to help you practice and hone your design skills. Many problems provide a set of technical specifications that can be met using the concepts and techniques covered in your engineering courses. For these problems, multiple solutions exist, requiring you to weigh the relative advantages and disadvantages of each design alternative. Although the principal focus of this book is design, it should not be relied upon as your sole design experience. If you are to learn to be a real engineer, it is crucial that you engage in actual design activities under the mentorship of experienced teachers, employers, or colleagues. You may find yourself confronted with design requirements in courses, projects, and assignments.

AN OVERVIEW OF THE DESIGN PROCESS

From a practical point of view, the process of designing a device or system involves several important steps. The first should be a definition of the overall design objectives. Although this task may seem trivial to the student eager to build and test real things, it is one of the most important. Only by first considering the "big picture" can an engineer determine all the factors relevant to the design effort. Good design involves more than making technical design choices. Key questions must be answered: Who will use the system being designed? What will it look like? What are the needs of the end user? What features are critical, and which are desirable but not crucial to the success of the project? Can it be easily manufactured? How much will it cost? Will it be safe?

To answer these questions, the designer must be familiar with the end user and with the environment in which the device or system will be used. Human factors such as physical appearance, ease of use, size and weight constraints, affordability, and safety must be considered along with such issues as method of construction, level of technology, and method of manufacture. These considerations should all share equal weight in technical design decisions.

A second step in the design process involves the selection of a design strategy. At this stage the engineer might decide, for example, whether the design will involve an electrical or mechanical solution, or whether it will be synthesized from "off-the-shelf" components or from basic raw materials. If the system is complicated, it should be broken up into simpler, smaller modules that can be interconnected to form the complete product. Modules should be designed so that they can be individually tested before the entire system is assembled. Organizing the task in this way simplifies synthesis, testing, and evaluation, and helps to subdivide the problem into several smaller tasks easily performed by one person. In a team design effort, the modular approach is essential.

The design strategy should also take into consideration the results of prior research and development. In many cases, solutions to similar problems, or perhaps to parts of the design, may already exist in commercial form. The wise engineer makes use of existing products that simplify the design task. There is no shame in using "off-the-shelf" components or subsystems as part of an overall design effort (provided, of course, that including them does not lead to patent infringement problems if the product is being developed for sale.) A good designer will always look for shortcuts to the final design objectives. Imagine how needlessly complex the task, for example, of designing a desktop computer without making use of disk drives, memory chips, power supplies, monitors, and central processors available from other vendors.

After the design strategy has been solidified, it is time to make a "first cut" at each module in the system. A detailed layout, engineering drawing, or other appropriate description of the design is formulated on paper and tentative values or dimensions assigned. At this stage, key quantities—sizes, dimensions, construction materials, stresses, electronic component values, for example—are estimated and evaluated. This step in the design process often involves rough approximations and gross estimates. Its primary purpose is to determine whether or not the design approach has a chance of working, and it should result in a tentative design for each module or portion of the system.

In the next sequence of the design process, the paper design is evaluated, built, tested, and—most importantly—evaluated again. The design should be revised as many times as is necessary. It is this revision process that constitutes the principal work of the engineer. A good engineer will review a design many times, often proceeding through numerous iterations until the best configuration is found. The process of evaluation can take many forms. Prototypes can be built in temporary form or fabricated from raw materials. Computer aided design tools such as AutoCAD™, SPICE, or Simulink™ can be used to simulate pieces of the system. Simulations can save time and expense by helping to identify fundamental design flaws *before* the product is actually built.

After the design process has converged on a probable solution, the design should be thoroughly tested and "debugged." Despite the usefulness of computer-aided design tools, there is simply no substitute for constructing a real physical prototype and using it to test the design. Performance should be assessed from many points of view. The effects of temperature, humidity, loading, and other environmental and human factors must all be taken into account. The design should be modified if problems are identified at any stage. If it is electronic, has moving parts, or will be subjected to cyclical loading, the finished product should be subjected to an extended "burn in" period to help identify any latent defects that might cause the device to fail after extended use in the field. Only after a comprehensive test period is the product ready to be put into actual service.

DOCUMENTATION

When designing an engineering system, good documentation is crucial. An engineer must keep careful records of all tests performed and design elements completed in an engineering logbook. It's a good strategy to write everything down, even if an item seems unimportant at the time. Documentation should be written in such a way that another engineer who is only slightly familiar with the project can repeat all work done by simply reading the logbook. Careful documentation will aid in the task of writing product literature and technical manuals should the design be destined for commercial sale. Above all, good documentation will provide the engineer with an overview of the design history and the key questions that were addressed during the design process.

CONCLUSION

This book is about design. Design is the process by which engineers build real things, and it is best learned by practice and participation. By reading the various chapters and doing some of the end-of-chapter exercises, you will be introduced to design while learning about what lies ahead in your future career as an engineer. Whether you are a senior or a first year student, *Engineering Design* will provide you with insight into the life of an engineer and help you to make informed and intelligent career choices at critical junctions of your education.

Engineering Design

1

So You've Found the Cafeteria: First Day on the Job

Monday, 7:30 AM...

With dreams of summer vacation still in her head, Sally Lew, recent Electrical Engineering graduate, headed down the parkway. *First day on the job,* she thought. *Am I nervous? No.* She turned down A-Street. *Yes, I'm nervous.* She found a small sign that read, "Xebec Research and Development Corp," and headed toward a two-story modern brick building. Her fears and anxieties were typical for a just-graduated engineering student: Was my training sufficient for this job? Will I remember everything I learned in my classes? Will the boss think I'm competent? *That's Zee-beck,* she reminded herself. *The company name, Xebec, is pronounced "Zee-beck" with a "zee" and long "ee." It's named after the word for an ancient three-masted sailing vessel.* The company president was into sailing.

Sally could still recall the anxiety that surrounded the first few weeks of her job search during her last semester at the university. At first, she wondered if engineering had been a mistake. Would all that hard work, time spent in classes, and long hours spent on assignments and projects relate at all to the real world? Would she find a job? She had signed up for a number of job interviews at the campus placement center and had even bought a new suit for the interviews. She had made the decision that she was going to land a good job and had gone to the interviews with a positive attitude.

Most of her friends from the dorm had been looking for jobs, too. As the only engineer on the floor, Sally had felt that her prospects were better than most, and had indeed been right. Katherine, her English major roommate, had been accepted to graduate school but was unable to find financial support. Kate had finally taken a job at the local bookstore in her home town. Mike, a psychology major from the next floor up, went to work for an architectural firm as a human relations trainee. Deb, a fine arts major from down the hall, was still looking for a job. Only her best friend Aileen, a business major she'd met in calculus class, had also found a job with a large company doing work that would fully utilize her college skills. Now Sally was in the same position. She had been hired over the summer by Xebec, a medium sized company of about five hundred employees specializing in prototype development, engineering design, materials research, and small government contracts. During her interview, Sally had learned a lot about the company and its many activities. *They're into all sorts of things,* she remembered thinking. Recent projects had included the design of door latches for the NASA space shuttle, vacuum chambers for Lanark Semiconductor, a video pipeline inspection system for East Coast Gas and Electric, and a computerized tool and die system for one of the "Big Three" automobile manufacturers. She had also learned about the company's small product division that specializes in surgical implants made using a patented surface hardening process. Before the Xebec process, artificial knee and hip joints lasted only five to ten years before wearing out their bearing surfaces. Xebec's patented surface hardening process produced joints with a projected lifetime of twenty years or more. Xebec, Sally had concluded, was a highly versatile, diverse company with expertise in many areas of interest to her.

An image of her first conversation with Harry Vigil, her future engineering manager, came to mind. Xebec had just received notice of an incoming contract to build the prototype for an electric automobile. A group of high-level investors, interested in marketing the car, had agreed to finance the project.

"Our investors feel they'll be able to compete with the Big Three," Harry had said. The "Big Three" meant Ford, Chrysler, and General Motors. "The market for electric cars is just beginning to emerge, and everyone, including us, is going to be starting out at the same place in the design process. We're a 'lean and mean' company, and if we approach the problem smartly, we'll be able to design a car that will beat our behemoth competitors in the marketplace. We're going to have to design a vehicle that's cheaper, more durable, and more efficient if we're to succeed. The biggest problem with electric cars," Harry had said, "is their efficiency, not their speed. Low efficiency means that stored battery power gets used up too quickly, reducing the driving distance between recharges. Compared to the energy content per weight of gasoline, the energy content of a fully charged battery is *much* smaller. Higher efficiency in the form of lighter weight and better control circuitry is the key to success."

Sally vividly remembered the next part of the interview. "Did you read *The Soul of a New Machine*?"[1] Harry had asked. Sally had looked up the book later that evening. It told the true story of how a project manager at Data General, a large computer company near Boston, Massachusetts, has assembled a cadre of junior engineers to build a state-of-the-art minicomputer in the late 1970s. "That's how I'm going to build this car—by using junior engineers who have enough training to get the job done but not so much they are biased toward 'the way we've always done things' or 'tried and true' designs. Most importantly, I want engineers who won't know enough to say 'we can't do that.' And I want engineers who won't be afraid to tackle a project with a low budget."

Harry had probably said too much about his strategy, Sally had thought at the time. In *The Soul of a New Machine,* the project manager from Data General had actually thought the assigned task to be impossible. He took a great chance at failure with his junior engineers, but in the end, unaware that the project was impossible, the junior engineers produced a winning design. The prospect of such a challenge had excited Sally, and she'd accepted the job. As she drove down the parkway, she found herself beyond excitement at having become part of the Xebec team. After a short summer vacation and a move out east, here she was, arriving for her first day at work.

Sally found a space in the Visitor section of the parking lot and began to get out of her car. *Wait,* she reminded herself. *I'm not interviewing anymore.* She was now an official employee of Xebec Research and Development. She moved her car to the more remote employee's section. It was still early, so she easily found a space.

Inside the lobby, a receptionist sat behind a desk typing into a console.

"Hi, I'm Sally Lew, here for new employee orientation."

"Oh, yes. Here's your temporary identification badge. Go down to the conference room, third door on the right." The receptionist checked off Sally's name on an alphabetical list. Sally looked at her badge which read

---

Xebec Research and Development
Temporary Identification/New Employee
Sally Lew
PD Division
No. 12856

Strange, thought Sally. *Strange to be wearing an ID badge and feeling like I'm being watched.* But she was no longer in the carefree environment of college. She knew that most companies protect their proprietary secrets with vigor. Identification badges and controlled access were more the norm than the exception at most companies. *Makes me feel a bit like a secret agent,* she mused. She clipped the badge to her lapel and headed down the corridor. The new briefcase she'd received for a graduation gift was under her arm.

Dave Jared, Mechanical Engineer, fidgeted with his tie. The suit felt strange. He was used to jeans and T-shirts with logos on them—the standard uniform of his dorm mates at State. Dave had joined Harry's engineering team as a new recruit after finishing up his B.S. degree the previous summer. He had attended summer term to make up the courses he missed when he was on co-op. Dave had been sitting in the conference room by himself since 7:30. He took the parkway bus to Xebec and it had arrived much earlier than he'd expected. *Better to be early than late the first day on the job,* he had noted. He'd helped himself to coffee from a fresh pot on the kitchenette counter in the conference room. Hopelessly addicted to caffeine during exam week his junior year, Dave wasn't sure he'd find coffee available when he took the job at Xebec. He'd even thought about bringing a thermos bottle. *That would be silly,* he'd decided. *The company* **must** *have a coffee pot. Any normal company would.* He had been right, of course. Not only did Xebec have coffee in abundance, they had a complete cafeteria in another part of the building. Dave had already found it in the half hour before the orientation meeting.

Harry had also lured Dave into Xebec with the promise of an extraordinary opportunity. "I'm assembling a team of four engineers," Harry had said, "fresh out of college and all crackerjacks in their fields: an electrical engineer, a mechanical engineer, a computer engineer, and an industrial engineer." Dave had not been sure about the last category. He'd asked around at school after the interview and learned that an industrial engineer studies and designs how things are made: manufacturing processes, materials usage, job queuing, finish work. It would be up to the industrial engineer, or IE, to make sure that the prototype car, as built, could eventually be manufactured on a mass production line.

"I think you'll fit well into this team as the mechanical engineer," Harry had said. "Your transcript looks good, but we've been most impressed by your hands-on experience. Tell me again about your senior design project. A robot, was it?"

"A robot *arm*," Dave had offered, "as an adaptive assist device for the handicapped. We worked on it in conjunction with a team of electrical engineers from the EECS Department."

"I remember now," said Harry. "You showed me an article about it from your home-town newspaper. Well, in any event, if you decide to take this job, you'll use those same mechanical skills when you become part of the Xebec electric car design team. Your principal activities will center around the design of the car frame and its associated mechanical structure, including the selection of the best materials for the job."

"Yes, I have a strong background in materials, loading, and testing as well," Dave had offered.

"Good. Each of the junior engineers I'm hiring will be assigned to a senior engineer who will act as your mentor and give you advice about your design activities. The senior engineers I've chosen have been with the company a long time and have a good sense about how things are done at Xebec. You'll work with your senior engineer on design concepts, feasibility studies, cost projections, testing, and so forth. But I want *fresh* ideas on this project, and I'll consider them to be your primary contribution to the team." Dave had been immediately convinced. "Sign me up," he remembered saying. And here he was, drinking coffee in Xebec's conference room. His head turned toward the door as someone walked into the room.

"Hi, I'm Sally. Sally Lew."

"Dave Jared. Are you part of the electric car design team also? Guess I'm the new mechanical engineering hire."

"Yes. And I'm the new EE. They've hired me to work on power electronics circuitry for the electric car project. My specialty area—at least what I did a lot of in college—is circuit design, both analog and digital."

"Me too. I mean, I'm going to be on the electric car project also. I'll work on pieces of the mechanical design," replied Dave, a bit distracted. "I graduated from State this summer. And you?"

"Me too—this summer, I mean. I graduated from West Coast University. Can you believe we're actually here?"

A person they both recognized as Harry walked into the room accompanied by two others who, from the looks of silent apprehension on their faces, were also junior engineers.

"Sally, Dave," began Harry, "I'd like you to meet Keith Stein and Anne Richards. Keith's going to be the industrial engineer on the electric car team, and Anne will work on computer software and hardware." Harry completed the introductions and asked everyone to sit around the conference table. "Your agenda this morning, team," he continued, "is to attend an orientation session, get signed into the company, and become formally indoctrinated. Someone from Human Resources will be here soon to start the session. This afternoon, at 2:00, we'll come back to this room, where you'll meet your senior engineers. You can have the rest of the day to get acquainted and settled. We'll meet here again tomorrow, 8:00 sharp. We'll formally kick off the electric car project at that time. And, by the way, welcome to Xebec!" Harry added the greeting with a last minute attempt at gusto, but it was obvious that he was slightly embarrassed at having not offered it first, before his other remarks. He was saved from the awkward moment when a well-dressed, middle-aged woman entered the room. She instantly commanded attention.

"Pat Harrington, Director of Human Resources," she said to the group. "Welcome to Xebec Research and Development." She took great care to pronounce "zeebeck" correctly. She turned on an overhead projector as she announced, "Here's a summary of our agenda for this morning. We've got a lot of ground to cover, so let's get right to work." She placed her introductory slide on the overhead. It was a summary of the morning's agenda:

XEBEC RESEARCH AND DEVELOPMENT CORPORATION
NEW EMPLOYEE ORIENTATION

- Orientation Schedule
- Overview of the Company
- Organizational Chart
- Research Summary
- Product Summary
- Explanation of Benefits

"Here's a summary of what I'll be covering this morning." She read each entry, pointing to it as she went down the list, then put up her second slide:

ORIENTATION SCHEDULE:

  8:00 Presentation

  9:00 Payroll and benefits

10:00 Company tour

12:00 Lunch with Xebec president

  1:00 Office and cubicle assignment

  2:00 Technical meeting with Harry Vigil

"Here's a synopsis of today's schedule," she added. Pat handed out a printed sheet that was a duplicate of her slide. "At 9:00, you're all to head down to Personnel to fill out payroll and benefits paperwork. At 10:00, Hank, my assistant from the engineering division, will give you a full tour of the company facilities so that you'll know where everything is. At noon, you're to meet in the company cafeteria, special events room, where the president will host you all for lunch. He still likes to meet all the new recruits on the first day. By the way, if you want a good conversation tip for the lunch table, remember that the boss is into sailing—that's how the company got its name."

I knew that, thought Keith. *Gather all information and cover all the bases before going to your first interview,* he remembered thinking at the time. He had looked up the word "xebec" in the dictionary before coming to his first interview:

Xe·bec (zē′bek) n. a small, three-masted ship having an overhanging bow and stern and both square and lateen sails; once common in the Mediterranean.

Pat Harrington continued with her presentation. "Our company's organized into three major divisions: contract research, prototype development, and product manufacturing." She put up the slide shown in Fig. 1.1 which explained the relationship between Xebec's several branches.

"We're a small company—about five hundred employees—so the organization chart is very simple. Each branch is run by a vice president who ultimately reports to the CEO. The CEO started the company in 1972 out of his basement. We've grown steadily since that time to what we are today. Xebec went public in 1986 when its stock shares became available for sale and were included in the NASDAQ index. The president still retains a major share of the company, of course. We moved into our present building in 1991."

Pat pointed to the first box on the chart. "Administration handles everything non-technical about the company, from hiring and promotion to building maintenance. Human Resources, my branch, is called 'Personnel' in some companies. We handle the mechanics of the hiring process, manage and distribute company benefits such as health insurance, retirement plans, and tuition remission, and offer training and development seminars. The Comptroller's office watches over the money. They keep track

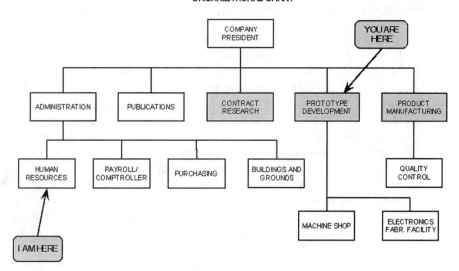

Figure 1.1 Fran's organizational chart for Xebec Corporation

of income and expenses, pay the bills, allocate funds, and manage the financial aspects of our contracts. Payroll, which is part of the Comptroller's office, will prepare your paychecks with the correct taxes deducted and make contributions to your 401K retirement plans. The Purchasing Department buys everything the company needs, from research materials and equipment to towels for the rest rooms. It's their job to get the best prices for everything and to process requisition forms. If you need something for your work, your department secretaries can show you how to fill out requisition forms, get the required signatures, and enter the proper codes to have the expense charged to your project budget. Buildings and Grounds, as you can probably figure out, takes care of our physical plant.

"Our Publications Department oversees every document that leaves the company. They check over and process project reports, instruction manuals, stockholder brochures, and proposals. They also can help you with the mechanics of document preparation—typing, layout, and printing—although these days most of our engineers prefer to generate their own documents and do their own writing on workstations or PCs. By the way, later this month a technical writer from Publications is going to lead you through a workshop designed to help you improve your writing skills.

"Now let me explain the three technical branches of our company." Pat pointed to the third box on the chart.

I already know all that from my interview, thought Anne as she prepared to tune out. *Why is she going over it again?*

"I know you all learned about the three branches during your interviews," continued Pat, "but you were all interviewed by different people. I want to make sure that everyone's up to speed with the same information."

That actually sounds like a good presentation technique, conceded Anne as she silently reprimanded herself and refocused on Pat's presentation. She was glad she had not protested aloud.

"The Contract Research Division, or CRD," Pat continued, "works on small and medium sized research contracts, mostly for various branches of the U.S. government. We've been rather successful over the years at winning contracts under the Small Business Innovative Research, or SBIR, program. Each year, the various branches of the federal government—mostly Department of Defense—publish 'wish lists' of research concepts they'd like to have developed or problems that need to be solved in areas of national need. Small companies such as Xebec can respond to the solicitations with proposals for pilot studies and initial development work. Success can lead to second- and third-phase contracts and eventual product development. We've also successfully bid on major research and development contracts to NSF, NIH, and NASA under the standard proposal route. Examples of recent R&D contracts include a new type of infrared sensor for night vision systems, a new coating to improve the flashover level of high-voltage insulators, and a new masking process for reducing the feature size of transistors on integrated circuits. All these projects, you may notice, have materials processing as a common denominator. Materials processing constitutes a large part of our contract research activity.

"Product Manufacturing is where we produce products on a large scale for public mass consumption. Many of our products resulted from initial R&D efforts in the Contract Research division. Examples of current products include surface hardened ball and socket joints for surgical hip and knee replacements, pre-processed silicon and gallium arsenide wafers, and non-contacting instrument probes for non-destructive testing applications. You'll get to see these production areas during your tour later this morning, including areas that I know you didn't see during your interviews.

"Now I'm sure you're all familiar with Prototype Development, or PD, the division under which you've all been hired, but let me review it anyway. About six years ago, in the face of diminishing federal dollars for contract R&D work, the president of Xebec started the PD division as a way to enhance company revenues and branch us off into new areas. Under the PD branch, we develop prototypes for engineering firms and other concerns that are too small to do the work themselves, or perhaps do not want to invest in engineering expertise to do something they've not done before or may be doing one time only. We begin with the initial product vision, design a prototype, then develop a complete concept for large scale production. When we're done, it's up to our clients to fund production, build the manufacturing facility, and actually produce the product if they so choose. It's a new concept, but of late we've had several major successes, including a new design for a recumbent bicycle and a video inspection system for visually detecting leaks in underground gas and water pipes. The electric car project, nicknamed 'Plugmobile' by the CEO, is our latest acquisition. A group of foreign investors has decided to enter the emerging electric car market with an independent design. As you know, you've all been hired to design it for them.

"Two facilities that fall under the Prototype Development division are the machine shop and the electronics fabrication facility. These folks will fabricate anything you need if you give them a proper set of blueprints. The machine shop can use computer generated CAD (computer aided design) files to run their computer controlled lathe and milling machines, and the electronics shop can work with your analog and digital circuit simulation files and printed wire board layout files to automate the circuit fabrication process. I'm sure that your senior mentors will show you the mechanics of getting designs to the fabrication stage when the time comes."

Pat talked some more about company organization, then went on to her last topic. "I'd like to end my presentation by reviewing the company's benefits." She put up the following slide, then explained the company policy on each item:

- Retirement Plan
- Health Insurance Plan
- Tuition Remission
- Vacation Policy

She talked at length about the various investment choices offered under the company's retirement plan. When she was finished, she offered an informal addendum. "Two pieces of advice, off the record. First: divert as much as you possibly can—between 5% and 10% of your salary—to your tax deferred retirement plan. I know that when you're in your twenties, retirement may seem like a long way off. But if you start now, you'll retire a millionaire. Run the numbers yourself. The wonders of compounding investment will *astound* you."

"Second piece of advice: I encourage each of you to take advantage of Xebec's tuition remission policy. The company will reimburse you for up to four tuition credits per regular semester at any accredited school or college you choose—there are several in this area—provided that the coursework leads to a higher academic degree or is related to company business. From what I've seen at most companies over the years, you'll never get out of entry level engineer status unless you earn a Master's degree, an M.B.A., or at least participate in continuing education."

Pat reviewed the health plan and vacation policies, asked for questions, fielded them, and then handed the floor over to Hank who had arrived during her discussion of benefits. The team followed Hank out of the room for the company tour.

"Not very much like college, is it Dave?" remarked Sally within earshot of Anne and Keith.

"I dunno, I feel so... grown up," said Dave with a shrug. He had offered the comment in jest, but then realized he was serious. "I'm earning money, having responsibilities, working on a major engineering project... thinking about retirement! Repaying my student loans!"

"Yeah," said Keith, "the scariest part is, this is real life now. We're out of the protected world of college. Although, I can't believe that I'm actually getting *paid* to work on this project. I'd work for free given the opportunity."

"Me, too," agreed Anne. "I think we're all in for quite an experience."

"I'm pleased you're all glad to be here," said Pat, overhearing.

"Let's get going, gang!" called Hank from down the corridor. "Tour time!" The four caught up with him to go on the company tour, eager to begin their lives as working engineers.

EXERCISES

1. Design your own engineering company. Pick a project of your choice and assemble a design team with the needed expertise.

2. Choose a real company about which you know something. Prepare an organizational chart describing the company and the function of each of its branches.

3. Find out about the health and retirement plans of your parents, guardian, relative, or older sibling. Write a short summary of these important benefits.

4. Calculate the value of a retirement investment consisting of a $2000 annual contribution for 40 years compounded monthly at 8% average return. Assume the dividends to be non-taxable.

5. Cite key factors in Pat's presentation that contributed to her successful technique.

6. Prepare a talk on a hobby activity for others in your class. Include background material, your involvement, and other accomplishments.

7. Suppose that you earn $35,000 per year as an entry level engineer. Look up and calculate the portion of your salary taken out for various taxes as well as company benefit deductions.

8. A product goes through many stages during its life cycle. In one view of the product life cycle,[2] the principal stages include concept, feasibility study, analysis, prototyping, design review, validation, production, sales, training and support, maintenance, and disposal. The following company employees are involved in one or more of these product stages: engineering manager, research scientist, electrical engineer, mechanical engineer, industrial engineer, machinist, sales manager, marketing director, lawyer, purchasing agent.

    Make a table in which the product stages represent the columns and the employees the rows. Try to envision the operation of an engineering company and check off the stages in which each employee is likely to be involved.

9. Interview someone you know who is an engineer. Write a short biography of that person, describing how he or she came to the decision to become an engineer. Describe the person's work environment and outline a typical workday.

10. According to one theory,[3] four types of individuals contribute to a team effort: the contributor, the collaborator, the communicator, and the challenger. The *contributor* is a dependable, task-oriented person who sets the standards for the team and makes sure that all available resources are used to their fullest. The typical contributor takes on the task of providing the team with technical information or parts data. The *collaborator* is a "behind the scenes" person who takes care of support details and strives to keep the team focused on its goal. The *communicator* seeks consensus among participants and encourages the involvement of all team members. Forever the peacemaker, the communicator attempts to resolve personality conflicts between team members. The *challenger* questions authority, embraces risk, and challenges the team's methods, goals, and outcomes.

    (a) Which personality type best describes you? Write a paragraph or short essay that justifies your response.

    (b) If you are presently part of an engineering design team, characterize each team member by one of the above personality types. Describe your choices in a one-page written paper.

    (c) Describe which of the above personality types best matches with the following traits: candid and open, commitment to goals, people person, always on time.

11. Choose a well-known engineer who is recognized as a pioneer in the field. Write a short biography of the individual, citing the factors that led to his or her engineering success.

12. Choose one device from the following list, then prepare a short (one- to two-page) paper that describes its technological evolution.

    (a) The personal computer

    (b) The automobile

    (c) Telephone communication

    (d) Supermarket checkout counter

    (e) Household lighting

    (f) Electronic music

    (g) Train transportation

    (h) Commercial building construction

REFERENCES:

1. Tracy Kidder, *The Soul of a New Machine*. New York: Avon Books, 1990.

2. Kim R. Fowler, *Electronic Instrument Design: Architecting for the Life Cycle*. New York: Oxford University Press, 1996, p. 32.

3. G. M. Parker, *Team Players and Teamwork: The New Competitive Strategy*. San Francisco: Jossey-Bass Publishers, 1991.

CHAPTER
2

Generating Solutions to Engineering Problems: The Art of Brainstorming

Tuesday, 7:30 AM . . .

It was a typical Tuesday. Sally headed into the company lot and parked her car. It had been about a week since the first day of work, and after the various meetings and technical design sessions that had been held, she was beginning to see the electric car project take shape. Her days were beginning to become routine now. She dropped her belongings in her cubicle, checked her e-mail (there were six messages waiting, none urgent; she would read them later after the 8:00 meeting), then headed for the coffee pot at the far side of the building. There she met Keith and Dave, already engaged in a discussion about what to expect at today's technical session. Technical sessions at Xebec were always interesting. Today's had the promise of being unique. Harry Vigil, their project manager, was about the lead the team through an idea trigger, or "brainstorming" session.

"Harry's memo said to come to today's meeting with an open mind and plenty of ideas," said Dave. "I was up half the night thinking about all the ways I could design the suspension system for minimum weight." After some thought, he decided to share what he was really thinking. "I hope my ideas impress the boss, or he's going to think he hired one real Bozo." Silently, the others shared his apprehension but said nothing. "Just kidding," said Dave, although everyone knew he really wasn't kidding. He changed the subject. "Anyone have any idea about what to expect? What's this brainstorming method that Harry said he would teach us today?"

"We learned about formal brainstorming in my senior design class last semester," said Sally. "It's rather neat, you'll see. It helps you tap the creative processes of your brain and emphasizes an important first step in good engineering design: how to generate ideas for subsequent evaluation. The brainstorming method also teaches you to not immediately settle on the first solution that comes to mind just because it's the first one, and not to settle on an approach just because it's the obvious choice. When I was working on my senior project, I learned the hard way that obvious is not always best." Sally recalled her experience during senior design class. She had insisted that her team design a temperature controller for a manufacturing process using an embedded microprocessor computer chip—to her it was the obvious engineering solution. Only after weeks of design work did the team realize that the problem could easily be solved using a few inexpensive analog operational amplifiers. The consequence of her poor choice had been a simple scolding and helpful discussion from her professor about the need to evaluate many design choices before choosing one. Had the setting been a real engineering company instead of school, her insistence on the first, obvious solution would have resulted in an expensive, needlessly complicated product.

It was now 8:00. Sally, Dave, and Keith left the coffee area, cups in hand, and headed to the conference room where Anne was waiting in the hallway. "How's the software wizard?" teased Dave as they entered the room. Anne curled her lip in jest and pretended to be annoyed, but was secretly pleased. She enjoyed the role of software expert that the team had placed on her. Always the quintessential computer hacker at school, she was glad to have found a job that tapped her computer addiction in a productive way.

The senior engineers were already seated, and all exchanged greetings. Harry came in with a sheaf of papers. As always, he spoke with animation. "Welcome everyone. Today's the big day when I teach you all about brainstorming. As we've discussed at previous meetings, each of you—Sally, Dave, Anne, and Keith—is responsible for designing a different aspect of the vehicle." Harry was repetitious as well as thorough. He would next remind them of their different responsibilities on the project. "For the near term, Sally, you are to continue working on the primary electrical system and power control circuitry. Anne, you should continue working closely with Sally and keep on top of designing the main computer control module and the software needed to run her power controller. Dave, I assume you're still working on the principal structural frame, suspension system, and body connections for the vehicle. And Keith, you're developing the best way to construct and manufacture the chassis and outer body. Have I got that right?" Everyone had learned to nod in approval each time Harry finished his overview of the labor assignments.

The hierarchy of responsibility was clear. At first, the senior engineers would carefully look over the shoulders of the junior engineers, checking design work and approving calculations and design choices. But as time progressed, Harry expected the junior engineers to take more and more responsibility for their design decisions. That's how they would ultimately earn their keep at Xebec.

Harry launched right in to the morning's agenda. "We're going to devote our technical meeting today to learning how an idea trigger session works. It's a structured form of brainstorming that we use often at Xebec. The idea trigger method is based on the work of psychologists and has been shown to enhance the brain's creative process.[1,2] It may seem contrived to you at first, but believe me, I've used it myself and it works. The technique will tap your creative potential and will help you to think of innovative design approaches that might not be obvious at first. All too often, we limit ourselves to the 'obvious' solutions or to whatever first comes to mind. Good design is more than that, however. Good design requires that we first consider as many alternatives as possible, discarding those that we deem unfeasible based on study, analysis, and engineering intuition. The design process also requires that we constantly evaluate, test, and redesign—possibly following several design approaches in parallel—until the results meet our requirements. Don't misunderstand me, I don't want you to exclude obvious textbook solutions as you work on the car project. I just want you to consider many options before choosing the final design path. Today's brainstorming session will help you to do that."

Harry passed out a printed sheet of paper to each participant. On it were four blank columns, as in Fig. 2.1. The senior mentors were familiar with the form. Each had been through the process before and lacked the curious puzzlement of the junior engineers. "To illustrate the brainstorming technique," said Harry, "we'll address the following problem: The car you're designing must be lightweight and as efficient as possible. In what ways can you design the vehicle to save weight and energy?

IDEA TRIGGER SESSION:
Contributor:

COLUMN 1	COLUMN 2	COLUMN 3	NOTES (or COLUMN 4)
2 Minutes			
30 Seconds			

Figure 2.1 Harry's entry sheet for the idea trigger session

"Now here's what I want you to do. Take two minutes, which I'll time, and write down in rapid succession under Column 1 as many ideas as possible. Key words will suffice; don't spend a lot of time writing whole sentences. This two-minute period is called the 'idea purge' phase. During this time period, I want you to open your minds, consider many alternatives, and don't worry if your ideas seem too trivial or ridiculous. Don't be afraid to write down 'pie-in-the-sky' types of ideas that may seem radical or impossible at first. In short, write down *anything* that comes to mind. Remember that everyone else will be under the same two-minute time pressure, so don't worry about impressing the others in your group.

"After the two-minute session, we'll take a two-minute break, then another thirty seconds of writing down ideas under Column 1. This 'tension and relaxation' sequence has been shown to enhance creativity. The idea is to completely purge your brain of all the ideas in its subconscious memory.

"After the idea purge phase, we'll proceed to the idea trigger phase. We'll go around the group, clockwise, and each of you, in turn, will recite all of your entries from Column 1. As you listen to others recite, silently cross duplicates off your list. And here's the important part: as other people mention their ideas, hearing them will trigger new ones in your own mind. Write these new ideas under Column 2 as soon as you think of them. This process is what actually constitutes the 'idea trigger' process. The remarks and ideas of others cause the hidden thoughts stored deep in your subconscious mind to surface. Remember, the purpose of the idea trigger phase is not to discount the ideas from Column 1, but rather to see if you can generate any new ones, modify your original ideas, or modify the ideas of others.

"After we've completed the first idea trigger phase, we'll go around the circle for another one. This time, you'll all read your Column 2 entries, writing down any new ideas that are triggered under Column 3. When we're all done, I'll compile everyone's sheets and make one master list of all the ideas that you've generated. At that point, we'll be able to discuss them, discard the ideas we think won't work, and decide which of the remaining ideas are appropriate for further consideration and development. And *that*, team, is the essence of how one begins the design process.

"Before we start, let me again review the ground rules for the session:

- No criticizing. We won't judge anyone's ideas until the final discussion phase at the end of the session.
- No hesitating. No idea is to be considered too outrageous, ridiculous, or 'wacky' to be entered under a column.
- No holding back because you think you have 'too many' ideas. The more the better.
- No owning of ideas. If you can build upon or improve someone else's idea, write it down.
- No sticking to your own field. Think globally across disciplines.

"That's it! I hope you'll learn to use this technique with each other whenever you're stuck on a problem, or whenever you just want to consider alternative approaches. Most importantly, let's see if we can apply it immediately to the electric car project."

The group nodded their understanding. *You've got to be kidding*, thought Anne silently. The process seemed foreign to the linear mind she had perfected as a computer engineer. But she resolved to give it a try and approach the process with an open mind. Sally and Dave were raring to go. Keith was neutral. Harry held up his watch. "Ready, set . . . go!"

The first two minutes passed quickly as people scribbled down their ideas. During the break, Sally cleaned her glasses, Dave rubbed his eyes and ran his hands through his hair, and Anne and Keith sat quietly and chatted. The final thirty seconds passed even more quickly.

When it was over, Harry said, "Time for the idea trigger phase." Keith led off the session by reading the entries from his first column:

IDEA PURGE PHASE

Column 1
_____

2 MINUTES:
- Support struts = plastic
- Use I-beams
- Composites
- Zinc air batteries (lightweight)
- Magnesium wheels = lightweight
- Honeycomb?
- Two seater = small

30 SECONDS:
- Built-in power cord
- Streamline the body

Sally listened to Keith and crossed out her duplicate entries. When Keith was finished, her first column, including crossouts, looked like this:

BRAIN PURGE PHASE

Column 1
_____

2 MINUTES:
- ~~No lead acid batteries (zinc AV)~~
- Higher voltage
  => smaller current for same power
  => smaller wires
  => less weight

- ~~Use composites for shell~~
- Switching power supplies for efficiency
- Solar charger (to supplement plugging into wall outlet)
- ~~Aerodynamic design~~

30 SECONDS:
- Optimize power flow through software
- Sense speed, bring up battery banks as needed
- Aluminum frame

Sally next read those of her entries that had not been duplicated by Keith. As Dave listened to Sally, an idea flashed into his head. *Flywheels! We can put a flywheel inside the car.* Dave reasoned that they could include a large flywheel inside the vehicle, then use it to brake the vehicle by converting the kinetic energy of the car's forward motion into stored rotational energy of the flywheel. When it was time to accelerate the car again, they could transfer the stored energy of the flywheel back into forward motion to help get the car moving again. The method might save a lot of battery energy and prevent the loss of energy as brake heat. The flywheel would also provide a gyroscopic action that would help stabilize the car during turns. Dave wrote down "flywheel" under his Column 2. *I'll need to review rotational bodies and stored energy,* he thought. *And I thought I'd never have to use my mechanics course again . . .*

The spoken trigger phase made its way around the circle. A great many ideas, some simple, some esoteric, crossed the table. When everyone was finished, Harry started the process again. As each person read their Column 2 entries, others wrote down new ideas under Column 3. As Dave read his flywheel entry from Column 2, Sally got another idea. She remembered learning something about using motors in the brake mode during her course on electric machines, which she had taken junior year as an elective. *When decelerating,* she thought, *we can use the car's kinetic energy to drive the motors as generators.* This process would transfer energy back into the batteries instead of wasting energy as heat in the brakes. The system would be much more efficient than using a flywheel and wouldn't have the added expense of the flywheel linking mechanism. Sally wrote down "brake mode" under her Column 3. *I'll have to review that electric machines textbook when I get home tonight,* she thought. *Good thing I didn't sell it back to the bookstore as a used book!*

The second idea trigger round progressed, and Harry started a third one. After about forty-five minutes, the entire brainstorming session was finished. "OK, let's take a break," said Harry, "then discuss some of these ideas." The team headed for coffee while Harry compiled the entries from the entire group. His combined list of entries from everyone's three columns looked like this:

COMPLETE LIST OF IDEAS FROM EVERYONE'S SHEETS
- Support struts = plastic
- Use I-beams
- Composites

- Zinc air batteries (lightweight)
- Magnesium wheels = lightweight
- Honeycomb?
- Two seater = small
- Built-in power cord
- Streamline the body
- Higher voltage
- Switching power supplies
- Solar charger
- Optimize power flow through software
- Sense speed, bring up battery banks as needed
- Aluminum frame
- Conventional brakes for backup assist/precision stop

After the break, Harry reconvened the team. "We'll now discuss our list of ideas, weeding out the ones that don't look promising. Let's compare, contrast, and combine the good ones, then select a few for further investigation or feasibility study."

Anne's senior mentor led the discussion. "What about composite materials—molded epoxy resins filled with carbon or glass fibers? Pros and cons compared to steel and aluminum, please?"

"Very lightweight, almost the same strength as steel," offered Dave.

"Easy to mold, difficult to paint and get a clean finish," added Dave's mentor.

"Aluminum is hard to weld, while composites can be glued or bonded with resin. But they're much more expensive than either aluminum or steel," added Anne.

The discussion continued for some time. When it was over, the group eliminated aluminum as the outer body material due to its lower mechanical strength, poor paint holding surface, and welding difficulties. They decided to let Dave, the materials expert on the team, look into the relative cost and manufacturability of carbon and glass fiber composites versus steel. Dave agreed to report back at the next design review meeting. The team also agreed that Dave and Sally, the mechanical and electrical engineers, respectively, would pursue the flywheel and motor/brake concepts for saving battery energy in more depth. They all left the meeting with lots to do.

Theory Behind the Idea Trigger Method

The idea trigger session is useful in group settings when many ideas must be produced for further evaluation and discussion. The method relies on a process of alternating tension and relaxation to tap the brain's creative potential. By listening to the ideas of others, receiving the foreign stimulus of other people's spoken ideas, and being forced to respond with counter ideas, a participant's habitual behavior patterns, personality traits, and narrow modes of thinking, which often serve as barriers that stifle creativity, can be momentarily broken, allowing ideas hidden in the recesses of the brain to come to the foreground. A participant who is shy, for example, and reluctant to offer seemingly "stupid" ideas will be more willing to do so under the alternating tension and relaxation of the idea-purge/idea-trigger sequence.

The entries that appear under the second and third columns (and the fourth column if the problem is complex) are usually the most creative. Such richness is thought to result from several factors. Often participants are secretly angered at having had their ideas "stolen" by others and are self-motivated to move on to new, unexplored territory. Simple competitive pressure can also propel a person toward new, original ideas. Conversely, seeing that one's ideas have not been duplicated by others can provide positive reenforcement, pushing the participant to come up with even better or more refined ideas. Some individuals may respond to their own non-duplicated entries with a desire to produce more as a way of "hoarding" the good ideas. Yet others may subconsciously think that augmenting previously discussed ideas fosters group cooperation and joint decision making. Whatever the psychological process, the idea-trigger method is known to work. The formal steps of the method, useful for groups of three or more, are summarized here.

Summary of the Idea Trigger Method

Idea Purge Phase (2-minute write/2-minute break/30-second write):

Write down as many ideas as you can think of under Column 1. Take a break, then finish your Column 1 entries.

Idea Trigger Phase (as much time as required):

Read your ideas, in turn, as others cross off duplicate entries from their lists and write new ideas under Column 2. When the group has finished, repeat the process, this time reading Column 2 entries and writing down any new ideas under Column 3. Repeat the process with a fourth column until all ideas are exhausted. Complex problems may require as many as five rounds or more.

Compilation Phase (as much time as required):

Combine entries from all lists and discuss them.

Other Brainstorming Techniques

As a practicing engineer, you may often find yourself in need of idea generation and brainstorming techniques to solve your engineering problems. The idea trigger method described above is useful, and it does work! Sometimes, however, when working on your own or with just one other person, or with a large group in need of an alternative method, it's equally effective to generate ideas using the classic brainstorming, or "deferred judgment" method.[3] In this method, minds go into spontaneous mode and, like the idea-trigger method, are absolved from traditional boundaries. The designated leader reviews the objectives of the session and writes down ideas as they are offered. As ideas emerge, the leader calls on raised hands, and participants state their ideas briefly and concisely. The flow of ideas proceeds spontaneously. One person should act as the recorder and write down all ideas as they are contributed.

The ground rules are the same as those for the idea trigger method: No one may criticize, solely own ideas, nor stick exclusively to his or her own field. Similarly, no idea is too outrageous, no idea is superfluous, and no idea may be discounted until the post-brainstorming discussion period.

When the group is large, additional rules are helpful. Should a person have an idea that is directly related to an idea just offered, he/she should say "priority" and be called upon next by the leader. This "piggy-back" method often results in the rapid progression of an idea from seed to fully developed concept. A good leader will also stimulate the session by offering topics for discussion during slack moments.

Brainstorming is an art and requires practice, but anyone can do it. You can brainstorm in a formal session, but can also do it while walking to school or work, riding the bus, driving a car, or sitting around the lunch table. You can even learn to brainstorm by yourself. Practice brainstorming on the problem statements provided below. Choose a partner, work alone, or work in small groups. See what new ideas you can generate!

BRAINSTORMING PROBLEMS AND DESIGN EXERCISES

Use brainstorming methods to generate solutions to the following problems:

1. You are given a barometer and a stop watch. In how many different ways can you determine the height of the Sears Tower in Chicago?

2. Design a basic sensing mechanism that can measure the speed of a bicycle.

3. Many international airline flights allow smoking in the rear seats of the aircraft. Design a system that will remove or deflect smoke from the front seats of the aircraft.

4. You are given an egg, some tape, and a bunch of drinking straws. Using these materials only, how can you prevent the egg from being broken when dropped from a height of six feet (two meters).

5. Devise as many different methods as you can for using your desktop computer to tell time.

6. Design a system for washing the outside surfaces of windows from inside a building.

7. Design a system to be used by a quadriplegic for turning the pages of a book.

8. Devise a system for automatically raising and lowering the flag at dawn and dusk each day.

9. Design a system that will automatically turn on a car's windshield wipers when needed.

10. Develop a device that can alert a blind person to the fact that water in a pot has boiled.

11. Devise a system for lining up screws on an assembly line conveyor belt so that they are all pointing in the same direction.

12. Develop a method for detecting leaks in surgical gloves during the manufacturing process.

13. Devise a method for creating a signal from a magnetic compass so that it can be interfaced with a computer running navigational software.

14. Given a coil of rope and eight poles, devise a method for building a temporary emergency shelter in the woods.

15. Devise an alarm system to prevent an office thief from stealing the memory chips from inside a personal computer.

16. Imagine custodial workers who are in the habit of yanking on the electric cords of vacuum cleaners to unplug them from the wall. Devise a system or device that will prevent damage to the plugs on the ends of cords.

17. Develop a system for automatically dispensing medication to an elderly person who has difficulty keeping track of schedules.

18. Develop a system for reminding a business executive about meetings and appointments. The executive is always on the go, but can carry a variety of portable devices and gadgets. Use your knowledge of existing communications systems.

19. Devise a system that will agitate and circulate the water in a small pond so that it will not become choked with algae. Assume that an electrical outlet is not available at the site of the pond.

20. Devise a system that will allow a truck driver to check tire air pressure without getting out of the vehicle.

21. For the purpose of generating new ideas, products are often grouped together by analogy. From the point of view of design, determine the common element possessed by each product in the following lists:[4]

> A stapler, a wire bottle brush, an incandescent light bulb.
>
> A revolving door, a venetian blind, a coffee vending machine.
>
> A fork lift, a pair of salad forks, a backhoe.
>
> A calculator, a flashlight, an electric coffee pot.
>
> An egg timer, a stopwatch, a traffic light.
>
> A bicycle, an automobile, a light rail vehicle.
>
> A bicycle, a human powered airplane, a rowboat.

22. Imagine that you are instructed by your doctor to take your temperature hourly after a bad accident. Both your hands are bandaged so that you cannot operate a thermometer. Use brainstorming methods with two or more students to see how many different ways you can take your temperature. Decide upon the most feasible method.

REFERENCES:

1. S. F. Love, *Mastery and Management of Time*. Englewood Cliffs, NJ: Prentice-Hall, Inc. 1981.

2. G. H. Muller, *The Idea Trigger Session Primer*. Ann Arbor, MI: A.I.R. Foundation, 1973.

3. A. F. Osborn, *Applied Imagination*. New York: Charles Scribner and Sons, 1979.

4. S. Pugh, *Total Design*. Wokingham, England: Addison-Wesley, 1991, pp. 251-258.

CHAPTER
3

Logbooks, Sketches, and Computer Aided Design

Wednesday morning . . .

Dave was in his cubicle thinking about a possible design for the front suspension system of the electric car. For the past several days, he had debated, both with himself and other members of the team, the pros and cons of rigid versus flexible suspension struts, spring-loaded versus torsion-bar suspension, and single versus double joints. The design team had long ago decided to go with composite materials—carbon and glass fibers bound up in epoxy resins—for the main structural frame. After some basic feasibility hand calculations and simple computer simulations of the various stress and strain loads, he was prepared to look at one promising design concept in more detail.

Dave took his engineering logbook off the shelf and opened it to a fresh page. The logbook was a standard, quadrille-lined (grid) white page notebook with a marbled black and white cardboard cover. Fixed to the front cover was a sticky label that uniquely identified the notebook and its contents. He had started this particular notebook on his second day at work. Its cover label read:

> Xebec Research and Development Corporation
> **LOGBOOK NO**: 1041
> **PROJECT:** Mechanical Designs for Electric Car Project
> **Covering the Period from** SEP 9, 1997 to _____
> **Principal Engineer**: Dave Jared

Naturally, Dave had left the "to" date blank; he would fill it in when the logbook was full. Per company policy, Dave would record in his logbook every idea, calculation, sketch, design, and piece of test data relating to the electric car project. Dave had not been surprised to learn that Xebec had a logbook procedure. Most engineering companies require the use of logbooks in some form as a means of providing project continuity and a permanent record of the design process.

The Xebec procedure was simple. Dave had signed out a blank notebook from the supply room and the records officer had given him a logbook number—1041 in this case. When his logbook was full, Dave would leave it on his shelf, then start another in the sequence after obtaining a new logbook number from the records officer. When the electric car project was completely finished, Dave's logbooks would be placed in the company archive and remain the property of Xebec. Should any future questions arise about what tests were done, who did them, what decisions were involved in the design work, or what calculations were performed, they could easily be answered.

As he reached for his logbook, Dave thought about another possibility. What would be the fate of his logbook should he ever leave Xebec for another job? He had no such plans—so far he loved the company and his job—but who knew what the future might hold? Some new engineer would read his logbook, no doubt wondering who this "Jared" fellow was. Dave conjured up a vision of the imaginary engineer poring over his entries, trying to make sense of them and wondering if "Dave Jared" had known what he was doing. This last thought made Dave resolve to be neat, accurate, and thorough with his logbook entries, and to write as if he were speaking to the future phantom engineer. And of course, there was always the matter of proving that Xebec was the first to think of an idea should patent or invention questions arise. He had a responsibility to the company and to the integrity of his job to keep a good logbook.

Dave found himself thinking about his lab courses at school. The teaching assistants had always let the students work on scratch paper. In fact, in one class, students had been encouraged to recopy relevant items from loose sheets into a nice neat notebook after the lab was finished. The notebooks were easier to grade that way. The doctored-up notebooks were submitted for grading at several points during the semester. Somehow this procedure, and the pretense that the notebook in any way resembled a running record of what went on in the lab, had always seemed like a charade to Dave. When he reached his senior capstone design course, everything was different. "Write down everything," his professor had said, "just like engineers do in industry. Keep a running record of what works, what doesn't, and what you've thought of but have not tried. Record every mechanical test, every calculation relevant to the project. I'll be grading you on engineering performance and on how well you can function in a real engineering environment, not on your ability to recopy lab notes." The professor had visited the lab in person at random times, demanding to see a student's lab notebook on the spot and noting if it were up to date and complete. Loose papers on the bench were a dead giveaway and received scornful reprimands. "Important information, perhaps vital to the project, could easily be lost," the professor had warned. "If you use loose papers for anything other than doodling, they don't belong in the lab." The senior design professor had not specified any particular format for the notebook, as his chemistry and physics professors had done. There had been no instruction to arrange entries in the "correct" format so as to please the professor. These lab notebooks were to be used as real engineering *tools*, and any format that met the needs of the student while serv-

ing as a well-organized record of the design history was deemed suitable. The procedure took some getting used to—stopping to write down everything instead of just
plowing ahead with the experiment or development work—but Dave eventually saw
the value of the notebook procedure when the time came to write up his final report.
All of the important information, including some he'd forgotten, had been entered into
the logbook and was at his fingertips. He was also able to explain why a previous iteration in the design had been abandoned. Dave was glad he learned about using logbooks and was well prepared to follow Xebec's logbook procedure.

Dave refocused on his work. It was time to record his new idea for the front suspension system in his logbook and to take a first cut at some calculations. He entered
the following sketch into his logbook:

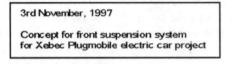

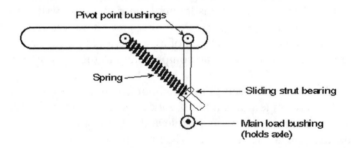

Figure 3.1   Concept for support strut for front suspension system

Dave was excited about his new design concept. It made use of a sliding strut bushing—
something that had never been tried in an automobile before. At least, he had not
been able to find any reference to the technique in the literature. The sliding bushing
would allow the horizontal component of the load force in the front end to be transferred to a spring, thereby reducing the transfer of jolts and bumps to the main chassis
and the operator. In theory, Dave reasoned, his idea would lead to a more comfortable
ride. In addition, the design would reduce shear stress, or sideways force, on the main
vertical strut. The carbon and glass composite material to be used in the Xebec car
frame structures had excellent strength in tension and compression, but less strength
in shear mode. If Dave's design worked, the team would be able to construct the front
end using composite struts of smaller cross-sectional area, and hence lighter weight.
Dave had gotten the idea for the sliding strut bushing while thinking about improve-
ments to his mountain bike.

He pondered the sketch, then entered a few trigonometric force calculations into
his logbook beneath the hand drawn figure. He was attempting to calculate the max-

imum expected force on the spring surrounding the diagonal support strut. Here are the entries that Dave placed in his logbook:

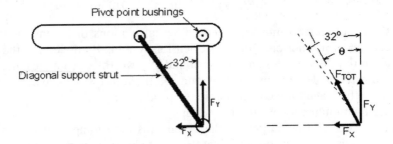

Figure 3.2 Force vectors on front suspension strut assembly

• Maximum expected vertical force at lower joint:

$$F_y = (2400 \text{ lb} \div 4) \times 4.45 \text{ N/lbf} \approx 2670 \text{ Newtons}$$

Based on 1/4 the design weight of the vehicle (there are four wheels) plus the full rated cargo/passenger load.

• Maximum expected horizontal force:

$$F_x = 2670 \text{ N} \times 0.6 \approx 1600 \text{ Newtons}$$

(based on assumed 0.6 g shock force.)

• Compute net magnitude of resultant force:

$$F_{tot} = (F_x^2 + F_y^2)^{1/2} = [(1600 \text{ N})^2 + (2670 \text{ N})^2]^{1/2} = 3113 \text{ N} \approx 3100 \text{ N}$$

• Compute angle of F_{TOT} relative to the vertical:

$$\theta = \tan^{-1} F_x/F_y = \tan^{-1} (1600 \text{ N})/(2670 \text{ N}) = 30.9° \approx 31°$$

• Angle between diagonal support strut and vertical is 32°. Close enough.

Dave next thought about the force transmitted to the diagonal strut. His idea was to surround it with a compression spring and attach it to the vertical strut with a sliding bushing. He wanted to compute the required properties of the compression spring. He wrote down some more calculations in his logbook:

• Reaction force F_R exerted by spring under compression. One end of the spring is fixed; x is the amount by which the other end is displaced:

$$F_R = -kx$$

Note: F_R acts in the direction opposite the displacement x.

Let's choose k so that spring displacement under maximum load conditions is limited to 1 cm, thought Dave to himself. He wrote some more:

• Compute desired spring constant for sliding strut compression spring. (The minus signs on displacement cancel):

$$k = F/(-x) = (-3100 \text{ N})/(-0.01 \text{ m}) = 310 \text{ kN/m}$$

Dave paused for a moment, checked his calculations, then continued with similar calculations for other parts of the suspension system. By the end of the morning, he decided that his new design had merit and was worthy of some testing. He would have the parts made from metal stock and use them to test his idea. Eventually, the parts would be made from composites in the production vehicle, but for initial mock-up testing, aluminum parts would do just fine.

The most novel part of the design was the sliding strut bushing. Dave drew a three-axis, multiview projection of the part to be machined, including dimensions, in his logbook. His sketch is shown in Fig. 3.3.

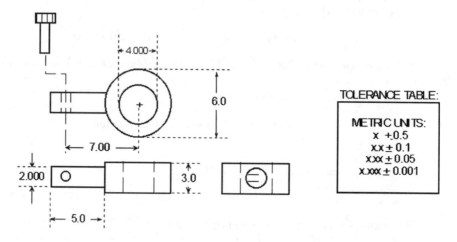

Figure 3.3   Three-axis projection of sliding strut bushing

Later that afternoon, he would make up a computer generated CAD version of the drawing. If the part made it to the final car design, the CAD file could be used to drive a computer-controlled lathe and milling machine on the production line.

Dave looked at the sketch of his part carefully and thought about how to specify its dimensions. No part can ever be machined to *exact* dimensions because machine tools leave imperfections and deviations in the machined materials. In preparing his drawing, Dave had to specify the acceptable deviation, or *tolerance,* for each of the bushing's various dimensions. As a rule, creating parts with tight tolerances involves the use of more expensive machining equipment and more machining time because material cuts must be made more slowly. These features add considerable expense to the finished part. As the designer, Dave had to decide which dimensions were truly critical. In his mind, only the mating surfaces of the bushing—the inner diameter of its main body hole and the outer diameter of its mating peg—were critical dimensions.

The format of the numbers on his drawing had precise meaning that the machinist would interpret by looking at the tolerance table. The outer diameter of the bushing, for example, needed to be 6 cm. The numbers 6.0, 6.00, and 6.000, though all mathematically equivalent, would mean different things to the machinist. Dave wrote the diameter as 6.0, with one digit after the decimal point. According to the tolerance table, this dimension should be interpreted by the machinist to mean 6 cm $\pm$ 0.1 cm. A bushing with a finished outer diameter anywhere between 6.1 cm and 5.9 cm would be deemed acceptable. Similarly, the location of the pin hole in the stud protruding from the side of the bushing was written as 7.00 on Dave's drawing, implying a machined tolerance of 7 cm $\pm$ 0.05 cm. The minimum and maximum tolerance limits for the hole location as machined would be between 7.05 cm and 6.95 cm from the center of the bushing. The most stringent dimensions of all were the diameter of the central hole through the body of the bushing and the outer diameter of the stud, dimensioned as 4.000 and 2.000, respectively, implying a strict machining tolerance of ± 0.001 cm for each.

As the morning approached lunchtime, Dave decided that he was pleased with his calculations and sketches. Following Xebec logbook procedures, Dave sought out Keith as a co-signer on his idea. The Xebec policy, typical for many companies, required that the logbook entry be signed by both the working engineer and a technically knowledgeable witness whenever a new, possibly patentable, idea was invented. Of the four junior engineers on the car team, Keith's background was closest to Dave's.

"Can you look over my logbook and be the co-signer?" he asked as he approached Keith's cubicle.

"Can I recheck your math?" asked Keith.

"I assumed you would," answered Dave. No answer was required. Keith had earned a reputation as a nitpicker with numbers and was going to check them over regardless of Dave's reply. He mentally reviewed Dave's calculations, doing "ballpark" estimates in his head. The numerical results all seemed to be in the right range. Dave's neat logbook entries and sketches allowed Keith to rapidly digest his colleague's concepts.

Keith decided that he was satisfied. "Seems like it might actually work," he noted. When Keith was finished reviewing his logbook, Dave added the following entry at the bottom of the page, and they both signed:

ENGINEER: <u>Dave Jared</u>   <u>**11/6/97**</u>
WITNESSED AND UNDERSTOOD: <u>Keith Stein</u>   <u>**11/6/97**</u>

The final logbook page appeared as in Fig. 3.4.

"Can we meet now to work on some CAD sketches at the computer?" Dave couldn't wait to use the company's computer aided design software.

"I'd love to do it with you," answered Keith, "but I'm tied up until this afternoon with technicians who are experimenting with a new painting concept for the production line. Can you wait until first thing after lunch? I'll bring dessert." Keith knew that Dave ran on his stomach.

"Sure," said Dave. "It's a deal."

28

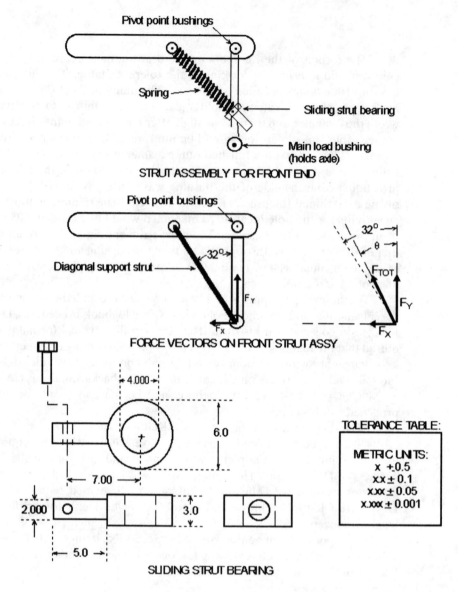

Pivot point bushings

Spring

Sliding strut bearing

Main load bushing
(holds axle)

STRUT ASSEMBLY FOR FRONT END

Pivot point bushings

Diagonal support strut

32°

F_Y

F_X

FORCE VECTORS ON FRONT STRUT ASSY

32°

θ

F_{TOT}

F_Y

F_X

4.000

6.0

7.00

2.000

3.0

5.0

SLIDING STRUT BEARING

TOLERANCE TABLE:

METRIC UNITS:
x ±0.5
xx ± 0.1
xxx ± 0.05
x.xxx ± 0.001

Engineer: __Dave Jared 11/6/97__
Witnessed and Understood: __Keith Stein 11/6/97__

Figure 3.4 Dave's complete logbook page

Later that day, 1:10 PM . . .

Keith and Dave were seated in Dave's cubicle. Both engineers had access to the company's CAD system over their UNIX-based workstations, but they decided to work at Dave's station because he had more room on his desktop. Dave's cubicle was also near a window—a precious commodity, even for an employee-conscious company like Xebec. They were both in the mood for sunlight, as it had rained the previous day. *Dave's luck of the draw to have a window,* thought Keith.

Dave loaded up the CAD software to his workstation from the resident server via the company network. Keith used the mouse to put up a rough outline of the bearing design on the system. Initially, they just outlined the basic shape of the bushing. Later they would enter more detailed features. When the entire CAD drawing was complete, they would print it out to send to the model shop machinist who was responsible for making one-of-a-kind parts for prototype testing. The finished CAD drawing would be much neater and much easier to understand than Dave's logbook entry. They would also send the CAD file over the network to an archive disk storage site where it would be retrieved at a later time should the need arise. If the bushing design was ultimately used in the finished car, the archived file would be copied to a manufacturing team for use in mass producing the actual parts from the chosen composite material.

Both engineers stared at the computer screen. "I suggest that we put a molded weld here, where the stud meets the bushing cylinder," said Dave. "That certainly would add the most strength." He pointed to the joint on the computer screen. "A composite weld is a strengthening bead added where two solid surfaces meet. In metal structures, the bead is added by welding after the individual parts have been machined. In composite structures, the beads are molded into the joint in the same location as if the structure were built from welded pieces of metal." The prototype aluminum part to be machined would be welded by hand. "Does the shop have heliarc welding?"

"Yes, it does," answered Keith. Heliarc welding was a process used to weld aluminum in the presence of helium gas. Aluminum cannot be welded in air because it oxidizes too rapidly, preventing bonding of materials at the weld site. "I agree with you about the welded joint, but if we include it all the way around, it will interfere with the pivoting action of the bushing on its own stud. I suggest that we machine a ridge in the peg right here . . . ," he ran his finger along the screen, ". . . and weld around that. It also will be much easier for the machinist to do the welding that way."

Keith and Dave worked on the CAD layout of the bearing design for about an hour. When they were finished, the screen appeared as in Fig. 3.5.

The drawings—top, front, and side views—would include all the relevant dimensions in both inches and centimeters. (The car would be built in metric, but the drawings would include both types of units for the benefit of the machinists who still liked to think in inches.) The log entry of text to be added to the drawing would describe the important mating surfaces and also include various notes and explanations that Keith and Dave felt necessary to communicate to the prototype shop.

"Guess that's it for now," said Dave. He moved and clicked the mouse and sent the file over the network to the printer at the machine shop. He then sent the following e-mail message from his login on the same workstation:

```
TO: DST
FROM: DAJ
RE: Bushing Design
```

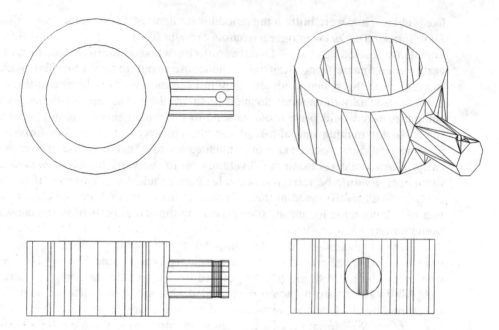

Figure 3.5   CAD drawing of sliding bushing design

```
Dale,
I've sent over to your printer the CAD drawing of the bushing we spoke
about yesterday. As we discussed, if you could get this part machined
sometime over the next few days, that would be great.

Thanks,
Dave Jared
```

Keith and Dave headed off toward the cafeteria talking about how they would set up the suspension strut test.

Later that day . . .

Keith was sitting in his cubicle, thinking about how Dave's sliding bushing and other parts for the support strut might best be manufactured. Although the strut design was by no means finalized, chances were good that the sliding strut concept would be adopted by the design team. It was important that Keith address manufacturing issues early on in the design process. It was his job to identify any parts that could not be easily manufactured or would add unnecessary cost to the vehicle. This evaluation process, called "design for manufacturability," was a relatively new concept in industrial and manufacturing engineering but one that Keith had learned well while an industrial engineering undergraduate. It refuted the age-old notion that manufacturing was an independent sequential event to be considered only after the technical specifications of the product had been finalized. This traditional method was sometimes called "over-the-wall" design by the old timers because design plans were thrown "over the wall" to a manufacturing team with the instructions, "Here, build this!" Most companies, including Xebec, now recognized that the dialog between manufacturing and engineering was an iterative feedback process. Design decisions had to go hand in hand with manufacturing decisions. Engineers on both ends of a finished product had to consider not only how it would function but also how it would be manufactured.

In specifying the manufacturing process for the electric car, Keith needed to consider such issues as assembly time, machining costs, materials costs, and handling time for each part, component, and sub-assembly. He now concentrated on Dave's sliding bushing, drew a sketch of the part in his logbook, and thought about how it might actually be made. *I could have it machined from one solid block of material,* Keith thought, *or made from two separate parts—the main sliding cylinder and the connecting stud.* He decided to compute the ultimate cost of the part for each manufacturing method. Under the sketch of the original, one-piece part, which replicated Dave's sketch of Fig. 3.3, Keith drew the alternative version shown in Fig. 3.6. The connecting

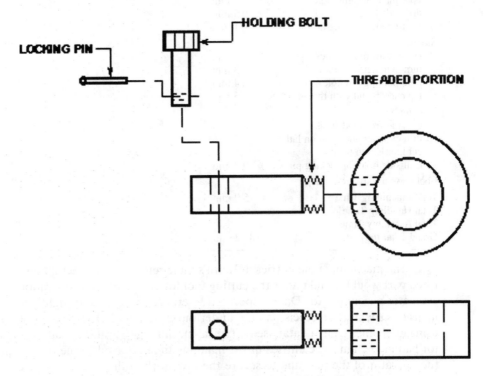

Figure 3.6   Keith's alternative two-piece sliding bushing design

stud now had a threaded portion which screwed into a tapped hole in the side of the main bushing. In machinist language, "tapping" was the process of adding threads to a hole, rod, or stud. It seemed like a reasonable way to manufacture the part. Keith also drew the locking pin which would be required to keep the holding bolt from falling out of the stud. He and Dave had omitted the locking pin from the earlier CAD drawings because the pin would be purchased as a ready-made part; it was not something to be sent to the machine shop for fabrication. The need for the locking pin, however, was something assumed and understood by both engineers.

Keith had decided that the part should be fabricated from high-grade steel stock. Although the main structural members of the front strut assembly would probably be made from composites, the friction bearing surfaces of the sliding bushings were best made from more wear-resistant steel materials. After contemplating the alternative drawing, Keith estimated the time cost associated with each part and entered the following table into his logbook:

Part	Machining Time	Manual Handling Time	Manual Ass'y Time
Main bushing		10 sec	120 sec
-cut disk from rod stock	0.2 min		
-turn cylinder on lathe	4 min		
-bore main bushing hole	1 min		
-drill side hole	0.5 min		
-tap threads into side hole	1 min		
Side stud		10 sec	30 sec
-cut stud from rod stock	0.2 min		
-turn cylinder on lathe	3 min		
-bore holding pin hole	0.4 min		
-tap end of stud with threads	1.5 min		
Holding bolt		5 sec	5 sec
-cut stud from rod stock	0.2 min		
-turn cylinder and head on lathe	2 min		
-cut knurls into head		0.5 min	
-drill side hole for locking pin	0.5 min		
Locking pin (buy pre-made)		5 sec	20 sec
Total Machining Time	15.0 min		
Total Handling Time		30 sec	
Total Assembly Time			175 sec
Grand Total for part:	18.42 min		

The machining time entries in Keith's table referred to the actual time the unfinished part would spend under the cutting tool for each particular operation. The time estimates accounted for Dave's specified tolerances, where more tightly specified dimensions spent more precision time under the cutting tool. Keith chose his estimates using guidelines from a database on typical machining operations that his senior mentor had given him. The entries under handling time referred to the time required for the operator of the machine to secure the raw stock in the chuck or vise of the computer controlled machine tool, then release the finished part from the machine and place it on a conveyor belt. Although the part itself would be cut by computer guided machine tools, the physical transfer of parts would likely remain a manual operation, at least under the manufacturing system that Keith planned to design. The manual assembly time entries reflected the time required to put together the parts on the car chassis. In general, many assembly operations on the car would be automated, but the assembly of the critical support strut structure would likely remain a manual operation.

Keith next revised the table under the assumption that the entire part (not including the locking bolt) could be made from one piece of material. The time required to machine the sliding bushing would be longer, but the assembly and handling times would be shorter because this method would eliminate the need to screw the stud into the bushing. He recorded the following revised estimates in his logbook:

Part	Machining Time	Manual Handling Time	Manual Ass'y Time
Main bushing/stud structure		10 sec	30 sec
-cut blank from block stock	2.0 min		
-trim rectangles into octagons	3.4 min		
-turn rectangles into cylinders	4.0 min		
-bore main bushing hole	1 min		
Holding bolt		5 sec	5 sec
-cut stud from rod stock	0.2 min		
-turn cylinder and head on lathe	2 min		
-cut knurls into head	0.5 min		
-drill side hole for locking pin	0.5 min		
Locking pin (buy pre-made)		5 sec	20 sec
Total Machining Time	13.6 min		
Total Handling Time		20 sec	
Total Assembly Time			55 sec
Grand total for part:	14.85 min		

As Keith's entries showed, the sliding bushing would require less overall time to make and assemble if it were made from a solid block of raw material rather than from two pieces. Since time translates directly into cost, more or less, the choice was clear—the part would be made from a solid block on the planned mass production line.

THE IMPORTANCE OF ENGINEERING LOGBOOKS - A SUMMARY

An engineering logbook becomes a permanent record of the design process, including all ideas, calculations, innovations, and test results. Logbooks become part of the company's archives at the end of the project period.

WHY AN ENGINEER'S LOGBOOK?

An engineer's logbook is intended to serve as a record of new ideas and engineering research achievements *whether or not they lead to commercial use*. The primary purpose for maintaining a complete logbook is that it will serve as evidence of inventorship, establish the date of conception and "reduction to practice" of a new idea, and show that the inventor has used diligence in advancing the invention to completion. In this respect, the Engineer's Logbook is more than just a simple lab notebook. It also serves as a valuable document with legal implications. The logbook becomes very useful should a key employee leave the company or be assigned to another project. Most engineering companies require their employees to keep up-to-date engineering logbooks.

PROCEDURE FOR LOGBOOK USE

Each company sets its own objectives regarding the keeping of engineering logbooks. The following guidelines, however, are typical of those specified by most companies:

1. Each engineer should keep a separate logbook. Notebooks with pre-numbered pages are recommended. All relevant data should be entered. When the logbook is full, it should be stored in a safe place until ready for archival storage.
2. All ideas, calculations, experiments, tests, mechanical sketches, flow charts, and circuit diagrams related to the project should be entered into the logbook. Entries should be in ink and should be dated. The entries should outline the problem addressed, tests performed, calculations made, etc., but subjective conclusions about the success of the tests should be avoided. The facts should speak for themselves.
3. The concluding page of each section should be signed and dated.
4. Important entries should be periodically and routinely witnessed by at least one other person, and sometimes two. Witnesses should endorse and date the relevant pages with the words, "Witnessed and understood."
5. Pages should not be left blank. If a page or portion of a page must be left blank, a vertical or slanted line should be drawn through it. Pages should not be torn out.
6. Relevant plots, graphics, schematics, or photos printed out on loose paper should be pasted or taped in as needed.

EXERCISES

1. Begin to keep a logbook for your class activities. Enter sketches and records of design assignments, inventions, and ideas.
2. Pretend that you are Alexander Graham Bell, the inventor of the telephone. Prepare several logbook pages which describe your invention.
3. Pretend that you are Marie Curie, the discoverer of the radioactive element radium. Prepare several logbook pages which describe the activities leading to your discovery.
4. Pretend that you are Dr. Zephram Cockran, the inventor of plasma warp drive on the television series *Star Trek*. Prepare several logbook pages which describe your invention.
5. Imagine that you were Dave designing the sliding strut bushing front suspension system. Create logbook pages describing the tests of the mock-up design on a bumpy road simulator. A set of data showing force on the diagonal strut spring versus time is shown below.

Time (s):	1	2	3	4	5	6	7	8	9	10
Force (N):	2000	2100	2700	2650	2810	1790	2220	2300	1930	2200

6. Imagine that you are Elias Howe, the first inventor to perfect the sewing machine by putting the eye of the needle in its tip. This innovation made possible the bobbin system still in use today. Prepare several logbook pages that describe your invention and its initial tests.
7. Reconstruct logbook pages as they might have appeared for the person inventing the common paper clip.
8. The accuracy of any number used in technical calculations is specified by the number of *significant figures* that it contains. A significant figure is any non-zero digit or any zero that does not serve to

locate the decimal point. A number cannot be interpreted as being any more accurate than its least significant digit. The numbers 128.1, 1.5, and 5.4, for example, imply numbers that have known accuracies of ±0.1, but the first is specified to four significant figures while the second and third are specified to only two. If trailing zeros are placed *after* the decimal point, they carry the weight of significant figures. Thus the number 1.000 means 1 ± 0.001. A quantity never should be specified with more digits than justifiable by its measured accuracy.

The accuracy of any computation can only be deemed as accurate as the *least* accurate number entering into the computation, and the number of significant figures that can be claimed for the result should be set accordingly. For example, the product 128.1 × 1.5 × 1.8 entered into a calculator produces the result 345.87, but because 1.5 and 1.8 are specified to only two significant figures, the rounded-off result of the multiplication must be recorded as 350, also with two significant figures. Note that a digit is rounded *up* if the digit to its right is 5 or more; if the digit to the right is less than 5, the trailing digits are dropped.

Evaluate each of the following numerical computations, expressing each result with the proper number of significant figures.

a. $F = 1221$ kg × 9.8 m/s^2

b. $V = 56$ A × 1200 ohms

c. $x = 76.8$ m/s × 1.000 seconds

d. $m = 56.1$ lbs + 45 lbs + 98.2 lbs

e. $i = 91.4$ V ÷ 1.0 kohms

f. $P = (5.1 \text{ V})^2/(1.0 \text{ kohm})$

9. When adding forces or other quantities represented as vectors, the principles of vector addition must be followed. Vectors to be added are first decomposed into their respective x, y, and z components. These components are added together separately, then recombined to form the total resultant vector. It is also sometimes convenient to decompose vectors to be added into components lying on axes other than the x, y, and z axes.

a. Two guy wires securing a radio antenna are connected to an eye bolt. One exerts a force of magnitude 3000 N at an angle of 10° to the vertical. The other exerts a force of 2000 N at an angle of 75° to the vertical. Find the magnitude and direction of the total force acting on the eye bolt.

b. A guy wire exerts a force on an eye bolt that is screwed into a wooden roof angled at 30° to the horizontal. The guy wire is inclined at 40° to the horizontal. If the eye bolt is rated at a maximum force of 1000 N perpendicular to the roof, how much tension can safely be applied to the guy wire?

c. A large helium-filled caricature balloon featured in a local parade is steadied by two ropes tied to its mid-point. One rope extending on one side of the balloon is inclined at 20° to the vertical. A second rope located on the other side of the balloon is inclined at an angle of 30° to the vertical. If the balloon has a buoyancy of 200 kN, what will be the tension in each of the ropes?

d. An eye bolt is fixed to a roof that is inclined at 45° to the x-axis. The eye bolt holds three guy wires inclined at 45°, 150°, and 195°, respectively measured clockwise from the x-axis in the x-y plane. These wires carry forces of 300 N, 400 N, and 225 N, respectively. What is the magnitude and direction of the total resultant force? What are the components of force measured perpendicular and parallel to the roof line?

4

The Importance of Estimation and Hand Calculations

Thursday . . .

It had been about two weeks since Sally began working on the vehicle's power controller. The design was slowly beginning to take shape. What started out as a standard textbook version of a power control circuit had been metamorphosed by ideas from the recent brainstorming session. The team had indeed decided to incorporate Sally's motor/generator braking concept into the latest version of the vehicle. The circuit that Sally was designing would form the heart of the power distribution system, transferring electrical power from the batteries to the motors during acceleration and from the motors back into the batteries during deceleration. In the latter mode, the motor would run backwards as a generator driven by the car wheels. As of their last design review meeting, the team had also decided to reject the flywheel concept as being too costly. Adding a flywheel system would require complicated mechanical couplings to transfer energy to and from the flywheel rotor. The power electronics needed to realize the motor/generator concept would be much less expensive and more reliable in the long run. The decision to implement the motor/generator concept put an extra burden on Sally. She now had to design an electronic power controller that allowed power flow in *both* directions. Her controller also had to interface with Anne's computer module.

The team had thought about designing a car with two motors—one for each wheel—to eliminate the need for a differential gearbox. In conventional automobiles, the differential gearbox compensates for the different rotational speeds of the driven wheels as the car makes a turn. A two-motor system could accomplish the same thing by sensing the turn and applying different voltages to the two motors. Such a system was deemed too risky, however. Given the infancy of electric cars, there simply was not enough field data to affirm the use of a two-motor drive train. The team had decided to stick to the proven technology of a conventional differential gearbox.

At a meeting the previous week with Anne, Sally and she had outlined the basic functional requirements of the power controller and its computer interface. Sally's controller would receive appropriate signals from Anne's computer module, then apply an appropriate voltage to the motor. Anne's computer module would need to monitor commands from the driver as well as signals from sensors placed throughout the vehicle. The motor would operate in one of four modes: In the acceleration mode, power would flow from the batteries to the wheels. In the braking mode, the wheels would send mechanical power to the motor, which would convert the mechanical power to electricity stored in the batteries. In the freewheeling mode, no power would flow to or from the motor, allowing the car to coast unpowered. In cruise mode, the car would travel at constant velocity, its motor receiving just enough power to overcome wind and friction losses.

Sally stared at a block diagram of the power control system that she had previously entered into her logbook. Her logbook page is shown in Fig. 4.1. It provided an overview of the accelerating and braking modes of the system that she and Anne must design. *Let's begin with the basics*, thought Sally to herself as she sat at her desk. *That's always a good place to start.* She picked up her logbook and her calculator. *How much power must the controller control?* She performed a simple calculation to compute the peak power flow during an acceleration event. Keith and Dave had estimated that the weight of the vehicle would be about 2400 lbs, or about 1090 kilograms. The team had decided to work in standard international (SI) units of kilograms, meters, and seconds so that everyone's calculations would match. That decision had been a big coup for Sally, since Dave had learned to work in English units of pounds and feet in college, as do most mechanical engineers. The investors hoped to license cars for manufacture in Europe and Japan, so staying metric and using SI units throughout seemed like a good idea.

In order to estimate the maximum power requirements for the motor controller, Sally had to review her basic physics. Power for accelerating the vehicle would ultimately come from the batteries. Given conservation of energy—electric power would be converted directly to mechanical power, less any losses, by the motor—computing the needed mechanical power flow to the wheels during maximum acceleration would allow her to estimate the required maximum electric power flow.

Force equals mass times acceleration, thought Sally as she recalled Newton's law of physics:

$$F = ma$$

Sally next remembered the formula that equates stored energy with the product of the force exerted on an object and the distance over which the force is exerted. *Energy equals force times distance:*

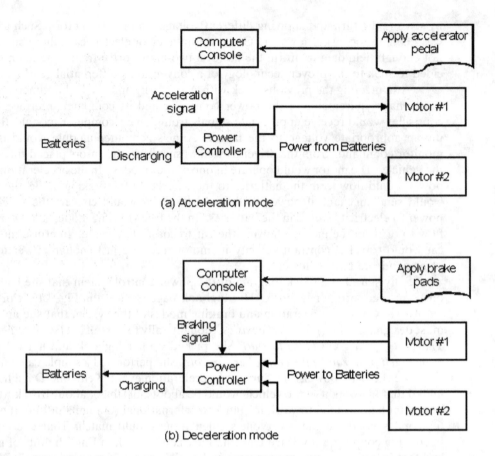

Figure 4.1 Block diagram of power control system and motor controller: (a) Acceleration mode (power from batteries via motor to wheels). (b) Deceleration mode (power from wheels via motor to batteries).

$$E = F \cdot x$$

Let's see, power equals energy per unit time, thought Sally. She wrote down the relationship between power and energy. The power would be equal to the flow of energy, measured in joules, per unit time:

$$P = E/t$$

The simple formula she entered into her logbook was valid for constant power flow. (For the more complex case of varying power flow, Sally could express the instantaneous power flow as the time derivative of the energy: $P = dE/dt$.) Sally assumed a desired maximum performance specification: from rest to 50 mph (about 22 m/s, or 80 km/hr) in thirty seconds. She then wrote down a few hand calculations in her logbook to determine the required acceleration:

velocity = acceleration $\times$ time:

$$V = at$$

acceleration = velocity ÷ time:

$$a = V/t = (22 \text{ m/s}) \div (30 \text{ sec}) = 0.73 \text{ m/s}^2$$

That's less than one-tenth the acceleration due to gravity, she noted, *9.8 m/s. Seems reasonable. As my favorite professor said, "Always think about your answer to make sure it's a reasonable number and is not ridiculous."* She continued writing in her logbook.
Force required to obtain maximum acceleration rate:

$$F = ma = (1090 \text{ kg}) (0.73 \text{ m/s}^2) \approx 800 \text{ N}$$

About eight hundred Newtons, thought Sally. *And at constant acceleration, the force will be constant also. Now let's compute the power required to sustain that force and rate of acceleration.* Sally reasoned that the needed power must continuously increase if her assumed constant rate of acceleration was to be realized. Her goal was to compute the peak value of power required. *Hmm ... time for some more physics.* Sally could compute the needed power in a few ways. She decided to try one based on her previously entered equations relating energy to total distance traveled. She assumed the vehicle to be initially at rest and to be accelerating at a constant rate beginning at $t = 0$. Constant acceleration would again imply a constant force. She could then express the energy acquired by the vehicle as the product of force times distance:

$$E = F \cdot x$$

Sally then remembered that for constant acceleration, the distance x would be equal to $at^2/2$ after time t. She wrote down the modified formula in her logbook:

$$E = F \cdot at^2/2 = ma^2t^2/2$$

She checked the units on the right-hand side of the equation just to be sure:

$$ma^2t^2 \text{ has units of } kg(m/sec^2)^2 \cdot sec^2 \equiv kg \cdot m^2/sec^2$$

A Newton is a $kg \cdot m/sec^2$, thought Sally, *so the expression indeed has units of Newtons times meters, which is energy.*
Sally next applied the formal definition of power as the time rate of change, or time derivative, of energy:

$$P = \frac{dE}{dt} = \frac{d}{dt} \frac{ma^2t^2}{2} = ma^2t$$

She assumed the 1090 kg car to accelerate at 0.73 m/s² for 30 seconds and computed the peak mechanical power input at the end of the acceleration period. This power would have to be transmitted to the wheels by the car's electric motor:

$$P = ma^2t = (1090 \text{ kg})(0.73 \text{ m/s}^2)^2(30 \text{ s}) = 17{,}425.8 \text{ Watts}$$

Oops! Sally remembered the admonition of her physics professor. *I'm not entitled to so many significant digits. I can only claim as many as my weakest estimate.* Because her least accurate factor had only two significant figures, she changed her answer to an approximate one of 17 kW. *The motors will need 17 kilowatts of peak mechanical power at the most demanding phase of the acceleration period,* she reasoned. *The motor will supply this mechanical power by taking in 17 kW of electrical power from the batteries.* Her power controller must be capable of handling a peak power of 17 kW.

Sally next focused on the electrical capacity requirements. The team had decided on a battery bank voltage of 96 V (eight 12-V automobile batteries in series). The electrical power going into the motor would have to equal the mechanical power transmitted to the wheels plus any electrical or mechanical losses in the motor. The motor would typically be 90% efficient. Sally drew the following power flow diagram, reflecting peak acceleration conditions, in her logbook:

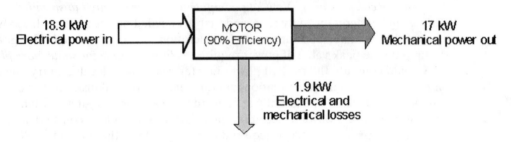

Figure 4.2 Motor power audit during peak acceleration

Electrical power = volts × current, $P = VI$, for a peak current of:

$$I = P/V = (18.9 \text{ kW}) \div (96 \text{ V}) \approx 197 \text{ A}$$

Sally pulled out her copy of *Handbook of Electrical Engineering*, a reference book she had bought during her senior year, and looked up the following wire table:

COPPER WIRE CURRENT CARRYING CAPACITY - AMPERES

Wire Size	Single Conductor-Open Air	Buried in Conduit
000	260	200
00	225	175
0	195	150
2	140	115
4	105	85
6	80	65
8	55	45
10	40	30
12	25	20
14	20	15

The table had two columns of values. Sally decided to use the less conservative "open-air" values since the cables would not be bundled in conduit under the car's hood. *We'll need wires of at least #0 gage.* In theory, a gage of #0 would be just thick enough.

A more conservative choice that included a safety margin for current carrying capacity would be the next higher size, #00 gage. However, the peak acceleration conditions would last for only a few seconds, Sally reasoned, and the wires could justifiably be sized according to the *average* power flow. Choosing #0 wire would save money in the long run and should be more than adequate—a reasonable design tradeoff. Number 0 gage wire could carry up to 195 Amperes and would be about 3/8 inch thick. *Not too bad,* thought Sally. *About the same as the battery cables in a conventional car.*

Questions about Sally's design estimates:

1. Will Sally's controller *always* supply 19 kW of power to the motor? What would be a reasonable estimate of the *average* power flow? Should Sally revise her wire diameter specification?

2. Sally's computations assume the motor to be 90% efficient in converting electrical power into mechanical power. All motors sustain losses, requiring more electrical power than their mechanical output. Revise Sally's estimates for power consumption and current flow if the motor is 95% efficient. Does this change alter the ampacity requirement of the wires?

3. Suppose that the performance specifications are changed to 0 to 50 mph in 45 seconds. Calculate the new peak power and wire ampacity requirements.

4. Suppose that smaller batteries are chosen that can supply only 100 A of peak current. Such a decision might be made to reduce battery weight and produce a lighter vehicle. If vehicle weight is reduced to 1800 kg by using smaller batteries, what maximum acceleration can be expected? What will be the peak battery current? What gage wire must be used?

5. Sally found the required instantaneous power flow directly from the relation $P = ma^2t$. Show that the physical units on the right-hand side of this equation equal energy per unit time, or power. Show that power also can be found by calculating the distance traveled over the elapsed time.

6. Why did Sally not choose #0 wires at 150 A capacity? Under what conditions would #0 wire work? What will happen if wires of insufficient current carrying capacity (e.g., too small a diameter) are used?

Later that evening . . .

The four junior engineers were out celebrating. Dave had proposed a gourmet dinner and offered his excuse. "It's been two weeks, and we're all still on the job."

"Never can tell with today's volatile job market," teased Anne.

"No, really—I think Harry is pleased with our work," said Keith. "Earlier this week he told me that the crop of new engineers was 'top notch' this year." It was clear that Keith was glad to be working at Xebec.

"My roommate who's a psych major is still job hunting," added Sally.

Spirits were high as Dave led them into Nick's Pizza Palace. "Gourmet? Your standards are distorted, Dave," said Anne.

Everyone took a seat, and Dave brought four sodas from the stand-up cooler. He planned to treat tonight—for the sodas, at least.

"You know, I was thinking," said Dave as he opened his Coke. "We could save the company a lot of money if we could invent a process by which different finish colors could be impregnated right into the body material during manufacturing." He often found himself thinking about the car project even after a long day at work.

"You've got to be kidding," Sally disagreed. "I can't believe the paint will cost that much compared to the rest of the car."

"It won't, for the paint itself. But labor is the main cost involved," submitted Keith. As the resident industrial engineer of the group, he was responsible for looking at all manufacturing processes (and defending his turf). "Run the numbers yourself— the time it takes to paint a car times the hourly pay of the worker, including benefits and overhead." The overhead at a typical fabrication facility—the cost of keeping the plant open so that workers could do their jobs—could run anywhere from 60% to 200% of the hourly wage.

"Unclear," said Sally. "Suppose the car is made by automated robots who do the painting. I saw a robot painting system during an I.E.E.E.[1] student tour last year," noted Sally. "Robots are commonly found on production lines, so your argument about labor cost really doesn't hold water."

The waiter arrived and set four glasses of water and four salads on the table. "Tell you what, then," said Dave, "just for fun, let's estimate the total cost of paint alone. How much paint will we need to cover the car? Cooperative estimation. Everyone in?" Two half-hearted nods and a weak "yeah" emanated from the group. Dave proceeded undeterred.

"No, seriously, if we compare the tooling costs of a robot system to the color impregnation equipment that I'm going to invent, we'll be able to see if there's any merit to my idea."

This last appeal convinced everyone. Anne offered to do the talking while they waited for the pizza to arrive. She pulled out a paper napkin. "OK, first thing we have to do is estimate the total surface area." She drew the rough sketch of Fig. 4.3 on the napkin :

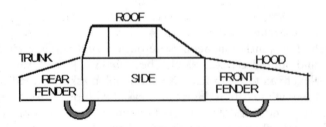

Figure 4.3   Anne's rough sketch of the car body

"We need to compute the area of each of the car's surfaces separately. The hood's easy—it's basically a 4 foot by 5 foot rectangle, for a total area of 20 square feet." She instinctively lapsed into English units of feet rather than metric, but nobody seemed to mind. "The trunk's also easy—another rectangle about 4 feet by 3 feet, for another 12 square feet. Umm... Let's model each of the sides by the following shape." She drew another sketch on the napkin, shown in Fig. 4.4. Anne pointed to the large rectangle. "This first shape represents the area *below* the window line. We don't need to count the surface area of the windows because they aren't going to be

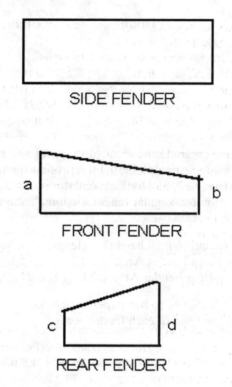

SIDE FENDER

FRONT FENDER

REAR FENDER

Figure 4.4 Approximate sections of car body

painted. The paint required to cover the posts and window frames is negligible. I figure that the length of the 'side fender' rectangle will be about 5 feet. The lengths of the front and rear fenders are about 5 ft. and 3 ft., respectively—the same as the lengths of the hood and trunk. Since it's only an estimate, we'll neglect the effect of the slanted hood and trunk on the dimensions."

She turned to Keith and Dave. "Guys, what are the numbers for a through d?" She pointed to the lettered dimensions on her paper napkin fender sketches.

Dave answered. "Umm... let's make a about two-and-a-half feet. Sound OK?" His stomach was beginning to growl and he'd become a bit distracted. "And b about a foot. That would make the d dimension 2.5 feet also, and let's say that c is a foot and a half. Pizza should have been here by now," he added.

"We forgot the roof," offered Keith. "About 6 by 5?"

Sally corrected him. "Can't be. If the length of Anne's side fender sketch is 5 feet long, and the windshield is angled steeply from hood to roof—you know, the side windows are more like trapezoids than rectangles—I'd put the length at the roof line at, oh... 4.5 feet. And about 4 feet wide—the same width as the hood." Everyone gave their approval.

"OK, I'll tally now," said Anne. She reached for a fresh napkin and made a list. She was being very methodical, as she was on the job:

ITEM:	LENGTH	WIDTH/HEIGHT	AREA
Hood	4 ft.	5 ft.	20 ft.^{2}
Trunk	4 ft.	3 ft.	12 ft.^{2}
Fender side	5 ft.	2.5 ft.	12.5 ft.^{2}
Fender front	5 ft.	2.5/1 ft.	(8.75 ft.^{2})
Fender rear	3 ft.	2.5/1.5 ft.	(4.5 ft.^{2})
Roof	4.5 ft.	4 ft.	18 ft.^{2}

"Now let me compute the areas of the front and rear fender sections [the numbers in parentheses in Anne's table]. I'll represent them both as the combination of a rectangle and a triangle." She took a calculator out of her purse. "We'll apply this formula to the two non-rectangular fender sections." She wrote down the following formula:

$$\text{Area} = (\text{length})(\text{small height}) + (\text{length})(\text{big height} - \text{small height})/2$$
$$\text{Front Fender Area: } (5 \text{ ft})[1 \text{ ft} + (1.5 \text{ ft})/2] = 8.75 \text{ ft}^2$$
$$\text{Rear Fender Area: } (3 \text{ ft})[1 \text{ ft} + (1 \text{ ft})/2] = 4.5 \text{ ft}^2$$

Anne added these entries to her paper napkin table. "Now I'll add up the columns. Remember there are two of each fender section.

$$\text{Total Area: Hood} + \text{Trunk} + 2(\text{Fenders}) + \text{Roof} =$$
$$20 \text{ ft}^2 + 12 \text{ ft}^2 + 2(12.5 \text{ ft}^2 + 8.75 \text{ ft}^2 + 4.5 \text{ ft}^2) + 18 \text{ ft}^2 = 101.5 \text{ ft}^2$$

"OK, let's call it about a hundred square feet. We'll work in round numbers since it's only an estimate," she said as the waiter placed the pizza in the middle of the table. Dave silently perked up.

Dave took a bite. "OK." He took another bite. "Now we have to come up with an estimate of the thickness of the paint. Any ideas?"

"How about a millimeter?" offered Sally.

"No way," said Dave. "I painted houses last summer, and I can tell you there's no way that paint is a millimeter thick. I'd say more like a mil (0.001 inches)." Everyone discussed the thickness estimate at great length.

Keith let them tighten the noose. He took a bite of his pizza, then offered the answer. "Well, you're all wrong. It just happens I've been studying ASME[2] standards as part of my production line quality control program. We need a paint layer thickness of 7 mils."

"Oh-kay ..." said Dave.

"All right," said Anne, "let's calculate the paint volume using Sir Keith's value of thickness." She grabbed a third clean napkin and picked up her pen between bites of pizza. She first converted the previously computed area to square inches:

$$\text{Area} = (100 \text{ ft}^2) \times (144 \text{ in}^2/\text{ft}^2) = 14{,}400 \text{ in}^2$$

She next multiplied the above by the 7-mil thickness of the paint to obtain the volume:

$$\text{Volume} = \text{area} \times \text{thickness} = (14{,}400 \text{ in}^2) \times (0.007 \text{ in}) = 100.8 \text{ in}^3$$

"About a hundred cubic inches. How many cubic inches in a gallon?"

"*I* dunno," said Dave. Keith shrugged. Sally squinted her eyes upward.

"Okay, I'll convert to metric." Anne multiplied the cubic inches times 2.54 cubed to get cubic centimeters.

$$\text{Volume} = (100 \text{ in}^3) \times (2.54 \text{ cm/in})^3 = 1639 \text{ cm}^3$$

"About 1600 cubic centimeters."

"Yup. And there's a thousand cubic centimeters per liter. So about 1.6 liters," noted Keith.

"Let's use the retail cost of paint, just to estimate," said Sally. "A gallon of best house paint is about $25, and a gallon is about 4 liters, more or less."

"So that's about $6 a liter, or about $10 worth of paint needed to cover the car," noted Keith. He multiplied 1.6 liters times $6 per liter in his head.

"But that's retail cost, and I used the cost estimate for latex house paint," added Sally. "I figure we're going to use an industrial type metal coating paint at wholesale prices. I'll bet the number is more like $6 per car." Everyone seemed to agree with her last minute fudge factor. "Our next challenge: The cost of a robot versus color impregnation equipment versus manual labor—amortized to a cost per vehicle in each case . . ."

Dave looked down at the table. "Hey, the pizza's all gone. Anyone want to order another?"

REFERENCES

1. The IEEE (Institute of Electrical and Electronic Engineers) is the primary professional organization representing electrical, computer, and systems engineers worldwide. It has student chapters on most campuses that hold a variety of activities, including seminars and tours.

2. ASME, the American Society of Mechanical Engineers, is the principal professional organization representing mechanical and industrial engineers worldwide.

ESTIMATING

Engineering design and estimation go hand in hand. When beginning any new design task, it's always a good idea to test its feasibility by doing rough calculations of important quantities and parameters. The exercises that follow can help you hone your design estimation skills. Discuss them with your friends, and see if you arrive at the same approximate answers.

PROBLEMS

1. Estimate the amount of paint required to cover a Boeing 747 airplane.

2. Estimate the cost of allowing a gasoline powered car engine to idle for 10 minutes.

3. Estimate the daily consumption of electrical energy by your dormitory, residence, apartment building, or home. (Check your estimate against real electric bills if any are available.)

4. Estimate the cost of leaving your computer running 24 hours per day, both with and without the monitor turned on.

5. Estimate the yearly energy cost savings if storm windows are installed on an average size four unit apartment building in your geographical area.

6. Estimate the gross weight of a loaded eighteen wheel tractor trailer.

7. Estimate the number of single family houses in your home state.

8. Estimate the number of bolts required to assemble the Golden Gate Bridge.

9. Estimate the number of bricks in an average size house chimney.

10. Estimate the number of piano tuners in Chicago.

11. Compute the total surface area of all the windows in your dorm, school building, apartment building, or house.

12. How much carpet would it take to cover the field in Yankee Stadium?

13. Estimate the total mass of air that passes through your lungs each day.

14. Estimate the time required for a stone to fall from sea level to the bottom of the lowest point of the Marianas trench in the Pacific Ocean.

15. How much does it cost to run a medium sized refrigerator for a year?

16. Estimate the weight of a single layer of shingles covering a wood-framed house with a pitched roof.

17. Estimate the physical length of a standard 120-minute VHS video tape.

18. How many coding pits are there on an average sized audio compact disk?

19. Estimate the number of books checked out of your school library each week.

20. When calculations are performed, the answer will only be as accurate as the weakest link in the chain. The answer should be expressed with the same number of significant figures as the least accurate factor. Express the result of each of the following computations with the correct number of significant figures.

   a. $V = (12.9 \text{ mA})(1500 \ \Omega)$

   b. $F = 2.69 \text{ kg} \times 9.8 \text{ m/s}^2$

   c. $F = -3.41 \text{ N/mm} \times 6.34 \text{ mm}$

   d. $i_B = (1.29 \text{ mA})/(100)$

   e. $Q = (6.891 \times 10^{-12} \text{ F})(2.34 \times 10^3 \text{ V})$

5

Computer Use in Engineering Design

Friday, 10 AM . . .

Sally and Anne were seated in Anne's cubicle discussing computers. Their main objective was to decide upon signal protocols for the interface between Sally's motor controller and Anne's computer module. Anne's module would serve as the main control system for the vehicle, integrating input from the human driver with numerous sensor signals. It would also produce the readouts for the car's dashboard display. The computer module would be built around an integrated circuit microprocessor, or "chip," and other peripheral components. The type of microprocessor, still to be chosen, would depend on the various requirements now being discussed by Anne and Sally.

A microprocessor, or computer brain, lies at the heart of just about every appliance or piece of equipment requiring intelligent control. Microprocessors can be found inside automobiles, microwave ovens, washing machines, children's toys, cellular telephones, fax machines, printers, and of course, personal computers. The Pentium™ chip made by Intel and the Power PC™ chip made jointly by Apple Computer and IBM are examples of high-performance microprocessors.

The microprocessor to be chosen by Anne and Sally would be a much simpler device than the Pentium or Power PC chip, but its operating principles would be the same. It would talk to the outside world via sets of terminals, called *ports*, designed to carry digital signals. A digital signal is a voltage that is either low (0 V), representing the binary number zero, or high (3 V or 5 V, depending on the system), representing the binary number one. A single port might consist of a group of 8, 16, or even 32 terminals activated by the microprocessor. A port is bi-directional—it can bring digital data into the microprocessor or send it out. The job of any microprocessor is to perform basic arithmetic or logical operations on data before sending the results back to the outside world. Microprocessors operate in base two arithmetic using a system of logic rules called *Boolean algebra*. The rules of Boolean algebra allow a microprocessor to make decisions based on the status of its incoming digital data.

"In order to control my power board," explained Sally, "I'll need individual one-bit signals that tell me whether to set the motor in *accelerate, freewheel, brake,* or *cruise* mode. Dave and I have decided, with Harry's approval, to handle *reverse* electronically, rather than via a gearbox, as in a conventional automobile. To reverse the car, we'll simply run the motor backwards. I'll also need a numerical value that tells me how much power to apply to the motors so that the driver's desired acceleration, communicated via the 'gas' pedal, actually occurs. I suggest that we express the desired acceleration, or applied power, in percent of total power available from the battery. I've done some calculations based on simple physics [see Chapter 4] that show acceleration to be directly related to applied power. I'd be happy to show you the calculations if you'd like."

"No, I get the idea," said Anne. "That relationship makes sense to me. Power is the derivative of energy versus time. Acceleration is change in velocity per unit time, and kinetic energy goes as velocity squared. I get it."

"Good," continued Sally. "I should also explain the constant velocity mode of the controller. We'd like to include a feature similar to 'cruise control' in this vehicle—one where the driver can press a button on a steering wheel lever and travel at constant speed until the next pressing of the brake pedal. On a conventional automobile, this feature is usually implemented by a separate control system involving hydraulics and mechanical linkage to the throttle. In our case, we can easily implement cruise control in software, since your computer control module will be constantly monitoring the car's speed. Your module will have to detect the first instance of brake action, of course, to get the car out of cruise control mode."

"Sounds like a 'can do,'" offered Anne, "but I'm glad we're going over it together. It helps me to reenforce my understanding of the overall system as we go along. One key question we need to answer is whether I'll send you the required percent power information in analog or digital form. If it's sent as an analog signal, then it will take the form of a voltage having a value anywhere between 0 V and 5 V, with 0 V representing zero power and 5 V representing full power."

"Right," agreed Sally. "My controller will send power to the motor in proportion to the ratio of the analog voltage signal received to a reference value of 5 V. In other words, a received voltage of 5 V will represent full power, use a voltage of zero will represent zero power."

Anne continued. "If the required percent power information is sent in digital form, then it should be sent as a binary number between 0 and 255, assuming an eight-bit data port."

"That's right," said Sally. "If the digital data port linking your module to my controller consists of eight terminals, or 'bits,' then the binary number it represents can lie anywhere between **0000 0000**, or zero, and **1111 1111**, which is 255 in base two. So, for example, if your module sends the binary number **1000 0000**, which is 128 in base two, that means it will be calling for 128/255th of the available power, or approximately half. If it sends **1111 1111**, then it will be calling for 255/255, or all of the available power."

"We're together on this," noted Anne. "Let's see . . . somewhere in the system we've got to do a conversion from the binary version of the percent power signal, which my microprocessor will compute, to the analog version of the signal ultimately required of your power electronics. I assume your thyristor board needs an analog signal?"

"It does," agreed Sally. "That will be the job of a circuit called the digital-to-analog, or D/A converter. A D/A converter takes a multibit binary signal as its input and produces an analog voltage as its output. Unlike a digital voltage, which has meaning only when it's high (logic **1**) or low (logic **0**), an analog voltage carries meaning for any value between its minimum and maximum limits. An analog voltage is what will actually control the power level of the thyristors in the motor controller board. We can put the D/A circuit either on your computer module or my motor controller board."

"Right, of course," noted Anne. "So, what's your opinion? Do we send the power control signal in analog form, so that the D/A circuit becomes part of the computer module, or in digital form, so that the D/A circuit is located on your motor controller?"

"Let's make a list of the tradeoffs," suggested Sally. She grabbed her logbook and made the following list:

POWER CONTROL SYSTEM

	Analog Signal Link	Digital Signal Link
No. of wires req'd	-two	-nine
D/A requirements	-may be able to locate inside μprocessor	-external design
Noise sources	-EM pickup	-digital "spikes"
	-contact resistance	-ignition noise
	-thermal noise	-digital crosstalk

"As a first consideration, let's think about the number of wires required between the computer console and motor controller." Sally continued to write as she spoke. "If we send the signal in analog form, performing the D/A conversion on your computer module, we'll need two wires, one for the analog voltage and one for the return ground. Two wires make a complete circuit."

"Yes," noted Anne, "but if we send the signal as eight digital bits, which is the standard port size available on many microprocessors, then we'll need nine wires—one for each of the eight digital bit (0 V or 5 V) signals plus a common return ground. In terms of simple wiring reliability, it would be better to send the signal as analog, since there are seven fewer wires. The types of connectors available for two-wire connections are much more robust than those available for nine-conductor digital wiring. The latter are typically flat, 'ribbon cable' connectors such as those found inside computers, whereas the former can be hefty, interlocking types of connectors such as those found in conventional cars."

"We have to consider many other factors, though, before making the final decision," suggested Sally. "Let's consider the D/A conversion operation itself. Some microprocessors contain the D/A circuitry right inside. Although the microprocessor chip itself would be more expensive, we'd save on overall chip count and peripheral components. If we send the signal as digital and have to perform the D/A operation on the motor controller side, all those chips and parts will have to be included on the motor controller board."

"True. So it seems like it would be best to send the signal as analog rather than digital," concluded Anne.

"Not so fast there, bubbie," joked Sally. "We haven't yet considered the most important tradeoff, and that's noise. My electronics professor spent a whole week on the subject of circuit noise, and I still remember the lectures. We came up with a saying: 'Good circuits and toys have low signal-to-noise.' The *signal-to-noise ratio* means the ratio of the desired signal to the unwanted noise. It must be high if the circuit (or toy) is to be a success."

Actually, Sally remembered the saying as it was *first* offered by her professor: *Low signal-to-noise separates the men from the boys.* When she reminded the professor that a third of the class was female, he had revised the saying, coming up with the alternative "toys" version. *We female engineers have to speak up,* she remembered telling him.

Sally refocused and continued. "Let's think about the various sources of unwanted electrical noise in our system." She continued to add to the list in her logbook. "Analog sources of noise include electromagnetic pickup from lights, appliances, power cables, and other devices operating at 60 Hz [50 Hz in Europe and most Asian countries]. All are sources of electric and magnetic fields. Analog signals will also be affected by changing resistance due to oxidation and dirt at the points where the signal wires are terminated by connectors. 'Thermal noise'—small bursts of current generated by random electron motion on a molecular level—will also be generated by the components themselves, but I think we can discount this thermal noise as being so small we don't need to worry about it."

"I should *say*," added Anne. She thought a moment, than said, "Hey, won't the car ignition itself be a source of electromagnetic noise?"

"Oops for you, Anne! There are no spark plugs or ignition wires. We're building an electric car—one that's driven by electric motors. But, in fact, the brush contacts bringing current to the rotating armature of the motor will generate considerable electromagnetic noise as the motor turns, which we should take into account."

"One thing to remember," noted Anne, "is that these sources of noise, which can add extraneous voltage components to an analog signal, can also affect digital signals."

"But in the latter case," explained Sally, "we only care about whether a given voltage is closer to 0 V or 5 V, representing the binary numbers zero and one, respectively. So the extraneous noise voltage components matter much less. In the analog case, these extraneous noise voltage components will be interpreted by the motor controller as signals."

"OK, you've made a case against analog signal transmission," recognized Anne, "but let's look at some of the noise sources if we send the power signal in digital form. If the physical circuit is not designed properly, the circuits producing those digital signals may generate digital ringing, or 'spikes,' caused by high-speed switching inside the microprocessor or other components."

"Yes, 'spikes' can occur whenever the wires carrying the signals behave more like capacitors and inductors than simple wires."

"We also have to contend with digital crosstalk—the tendency of digital signals on one wire to couple to other, nearby wires. Crosstalk can occur when a digital voltage undergoes a transition between low and high. Also, we need to consider . . ."

Sally cut her off. "The bottom line is, we can design the circuit properly to minimize all the sources of digital noise that you've mentioned. We can't as easily guarantee the noise integrity of an analog system. I vote for sending the signals in digital form, even though we'll need more wiring."

"Agreed, digital it is," said Anne. Sally had won her over. "One more thing we haven't considered, though. We could send the digital data in *parallel* form, all eight bits at once over eight separate wires, or in *serial* form, one bit at a time over *one* digital wire. (An additional ground wire would be required in both cases, of course.) Sending data serially would greatly simplify the wiring requirements. If we send the signal in serial form, timers on both sides of the data link will keep track of when the bits are supposed to arrive. If they time themselves independently, the link will be an *asynchronous* link. If common bit timing is maintained via a timing signal sent over a third wire, the link will be *synchronous*. The principal tradeoff is that serial links are usually much slower than parallel links. Similarly, asynchronous serial links are usually slower than synchronous serial links because extra time is needed to manage all the bit timing. Since our application is not particularly high speed, an asynchronous serial link should, in principle, work just fine."

"Also, many microprocessors contain the circuitry needed to convert the eight bits of parallel digital data into serial form, ready to transmit. So only the motor board will require the extra chips needed to convert the serial signal back to eight parallel bits ready for D/A conversion."

Anne and Sally continued their discussion of design tradeoffs; they weighed and measured the various advantages and disadvantages of each design choice. They finally decided upon an asynchronous serial link of digital data with D/A conversion performed on the motor controller board. This choice, they felt, was best given their desire for connection reliability and noise immunity. Signal transmission speed would be sacrificed somewhat, but they thought speed to be the least important design objective.

"Know what, Anne?" remarked Sally when they were done discussing signal transfer issues. "We also have to tell the motor the mode in which it must operate: acceleration mode for increasing speed, cruise mode for maintaining constant velocity, freewheel mode for coasting, and generator mode for braking. During acceleration, the appropriate amount of power must be sent to the motors so that the vehicle velocity will increase. But in cruise mode, the power sent to the motors should be just enough to overcome losses while maintaining constant car velocity. The freewheel mode, during which *no* power is sent to the motors, is needed so that the car can go into a coast, for example when the driver disengages the clutch. During brake mode, the motors must be run as generators, taking power from the wheels, thereby helping to slow the vehicle while storing the extracted energy in the batteries. We've already discussed the issue of the percent maximum power signal sent during acceleration. Cruise mode is a bit more tricky, because what we really want the system to do is maintain a constant velocity. If we shut off power completely while in cruise mode, the car will slow down due to the friction and wind resistance. Brake mode is no problem— the way I'm designing the power controller, the system will take care of itself as long as it knows that it's in brake/generator mode."

"I understand what we'll need to do," replied Anne. "During cruise mode, we need to monitor the velocity of the car and send that information to a feedback control loop in the software. The system will adjust the power sent to the motors so that the velocity stays constant. If the car starts to slow down in cruise mode, a bit more power will be sent to the motors. Conversely, if the car begins to speed up while in cruise mode, we'll slightly cut back on the power until frictional losses cause the velocity to go back to the desired constant value. That's the essence of a feedback control system."

"Of course, we'll have to analyze that feedback system to make sure it's stable," noted Sally. "If we attempt to correct the velocity too quickly during cruise mode, the car may overreact and the actual velocity may oscillate up and down, never quite settling on the desired value."

"True," said Anne. "I remember learning about the stability issue in my control theory class—the one that I took junior year. I'll review my textbook and look into the stability issue."

"Good," said Sally. "We'll temporarily set aside the feedback issue for now. But let's still talk about the four additional control signals required to tell the controller what mode it's in. We also need to tell the controller when to go into reverse, so that's actually a fifth state. I propose adding three additional lines between your console and my motor controller. They can share the same ground as the serial line, so that we won't need an additional ground wire. We can set each of the two signal lines to be either high or low, giving us a total of four binary states: **00**, **01**, **10**, and **11**. These four states can be used to signal cruise, accelerate, freewheel, and brake modes to the motor controller. The third line will be high (**1**) for forward and low (**0**) for reverse."

"Sure, that's one way we can do it," agreed Anne, "but we can also attach the 'mode' signal to the end of the power signal right over the serial line. We can simply send three additional bits of data either before or after the eight bits representing the percent power signal and use them, as you've suggested, to represent the four motor states, plus reverse. Make sense?"

Sally acknowledged the concept, but disagreed. "The bottom line, though, is that the motor mode signal is a crucial one—much more 'system critical' than the percent power value. If the motor goes into the wrong mode while the car is running, the safety of the occupants will surely be compromised. Can you imagine what would happen if the car accelerated when it was supposed to be braking, or went into brake or reverse mode on the freeway? In order to send the signal serially, we would have to make sure that the computer console and motor control board were always in 'sync,' and while synchronization of serial data may be 99% effective, we need better odds, like 99.99% or more, to ensure adequate safety. I vote for separate additional data lines for sending the motor mode information."

This last argument convinced Anne. She and Sally decided to include the three additional lines. "Give me some time to work on the flowchart for the program," suggested Anne. "Can we meet later to talk about it? I'd really like to get your input." The two agreed to meet sometime that afternoon.

Later that day . . .

Anne sat in her cubicle and reviewed the requirements for the computer software. The microprocessor, which would operate the entire console, would have access to several inputs. Binary signals—either "true" or "false"—would indicate the status of the ignition key as well as the accelerator, brake, and clutch pedals. Analog signals, converted to eight-bit binary numbers by analog-to-digital (A/D) converter circuits, would provide information about the car's speed and degree of accelerator pedal depression. The program inside the microprocessor would have to digest the available information and determine whether to accelerate, brake, cruise, or freewheel the motors. Anne sketched out the flowchart diagram shown in Fig. 5.1 in her logbook.

Every time the car's "ignition" (i.e., start key) was turned on, the power would be energized and the microprocessor would enter and run its own internal initialization, or "wake-up," program. The initialization program, specified by Anne in assembly code and stored permanently in the microprocessor's ROM (read only) memory, would function somewhat like the "boot" program on her desktop computer. It would prepare the processor's various registers and ports for input or output, as needed, and assign various program flags to be used by the main software program.

After the startup process, which would take no more than a few milliseconds, the microprocessor would begin executing its main software program as described by Anne's flowchart. The program would first query the various pedals to determine their status. If the brake was depressed and the car velocity zero, no power would be applied to the motor and the car would remain at rest. If the brake was off, the program would next check for reverse mode, set the motor direction accordingly, then proceed to test the accelerator pedal. If the accelerator was not depressed, the processor would next determine if cruise or freewheel mode was desired. If the cruise button was not engaged, the program would assume that freewheel mode was desired by the driver and completely disconnect the motor from the battery via the thyristor switch network, thus providing no power to the wheels. If the cruise button was engaged, the program would check the speed, record its value temporarily in RAM (random access memory), and return to the pedal check point. If the pedal status did not change, the program would again loop to the cruise mode branch, where it would recheck the speed and compare it to the stored value. If a decrease in real speed had occurred due to frictional losses, the motor power would be increased slightly before the program returned to the pedal check point. Conversely, if the speed had increased since the last

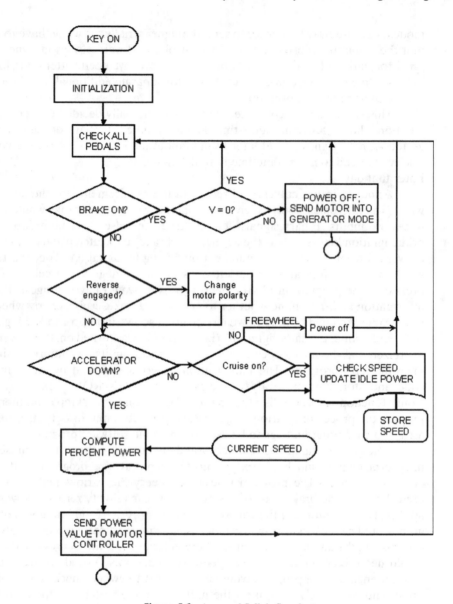

Figure 5.1 Anne and Sally's flowchart

cruise speed check, the power would be slightly reduced. Over many passes of the loop, each of which would take only a fraction of a second, the car speed would, on average, remain constant while the car was in cruise mode.

If the pedal check revealed a released brake and a depressed accelerator, the program would enter accelerate mode. It would assess the current speed, then compute the needed power to achieve an acceleration proportional to the amount of pedal depression before looping back to the pedal check point.

Anne was scrutinizing her own flowchart when Sally popped her head into the cubicle. "How's it coming along?"

"Not bad. I was hoping you could look over my flowchart and help me figure out if anything's missing."

Sally examined the flowchart at length, silently mimicking the logical flow with her hands as she mentally executed the program. "It seems to make sense. I assume for clarity of the diagram that you've left out a few of the logic steps needed by the cruise loop part of the program."

"That's right," said Anne. "I'll need to determine whether the cruise loop is being executed for the first time after any change in pedal status. The main task of the cruise loop is to compare present velocity with previous velocity, but if there isn't a previous velocity, the program must simply pass on through the loop. I can set a 'first pass' flag at startup, then clear it after the first pass of the program through the cruise loop. Any change in pedal status from cruise mode will reset the flag."

"Good," said Sally. "One other thing I would add. I should think that we'd want the microprocessor to do a full system check of the entire car immediately after initial startup, before the very first pedal check operation. When the key is first turned on, we could have the microprocessor check its own computer system, make sure all the sensors are operating, look for faults in the brakes or drive train, and even detect burned out headlights. The number of checks is limited only by the number of sensors we're willing to install around the vehicle."

"Great idea—Harry will like that one." Anne modified her flowchart so that the top section appeared as in Fig. 5.2. The two engineers studied the flowchart a while longer, then decided that Anne was ready to start writing some code.

"Sally," said Anne, "Could you give me your opinion about the best programming environment in which to generate the microprocessor code?"

Anne was referring to the fundamental language of microprocessors: a set of machine level instructions called *assembly code*. Used in some form by all microprocessors, assembly code language includes such low-level instructions as *read port*, *add binary*, *shift bits left*, and *compare byte*. Assembly language also permits simple decision making and conditional branching. More complex operations, such as the floating point calculations, data manipulation, and data communication required of the automobile console, can be performed by piecing together, or assembling, sets of the more basic instructions.

Anne's question centered around the following issue: Simple programs can be created quickly by writing code directly in assembly language. More complex programs are typically written in a higher-level programming language such as C, C++, PASCAL, or even BASIC. A special program called a *cross-assembler* reads the high-level program and translates it into appropriate assembly level code. Another program called a *compiler* then translates the assembly code into the basic binary machine language of the microprocessor. After compilation, the machine level assembly code

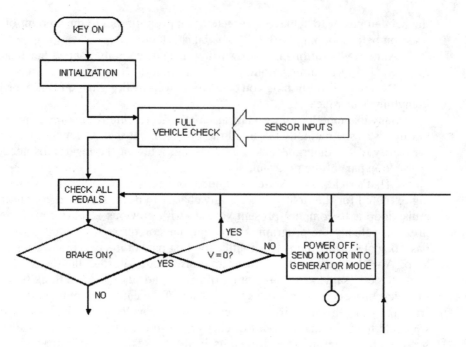

Figure 5.2 Modification to flowchart

instructions are transferred, or *downloaded*, to the microprocessor and stored in its permanent, read only memory (ROM). In the design stage, when a program is being developed, a special, more expensive version of microprocessor is used which permits the code to be erased and rewritten. The erase procedure requires that the micro-processor be exposed to ultraviolet light for several minutes. The chip package con-tains a special transparent window for this purpose. When the final design of the computer circuit is mass produced, less expensive versions of the microprocessor that permit a one-time burn-in only of the assembly code into ROM are used.

"I can't decide which high-level language to use for writing my program instruc-tions," explained Anne. "The C++ I learned in my programming course is certainly appropriate. Many cross-compilers are available for translating the program into mi-croprocessor assembly code. But I can imagine easily getting lost in the C++ code it-self because the language is vast and complex. There's also MATLAB™, which we learned during my first year at school. MATLAB's one of those mathematical calcu-lation packages that allows you to do complex tasks with simple instructions. There's good old FORTRAN, of course, which has been around for decades and with which some of the more senior engineers might be more comfortable. I've even thought about writing the program in BASIC, which I learned in high school and is fast, direct, and intuitively easy to understand, or maybe PASCAL. Any opinions?"

"Yes, several," replied Sally. "As I understand it, the strength of MATLAB is its ability to manipulate large amounts of data while allowing you to easily plot and graph it. You only have a few data variables—under ten, I'd estimate, even with all the sensors—and you really don't need to graph or visualize anything. Also, MATLAB ultimately executes its commands in C or C++, so generating the microprocessor assembly code is going to take an extra, possibly large, step.

"I don't think there are very many microprocessors out there for which a cross-assembler for PASCAL is available, so I'd rule out PASCAL. You *could* write the program in BASIC, and we'd probably still be able to find a decent cross-assembler. But BASIC is not an object-oriented programming language. The code you write is likely to be read and modified by many people in the future. For that we need a language that enforces the use of modular programming in which each section of code is linked to other sections of code only by its input and output variables. In BASIC, you can jump to any line of any module in the program at will, opening up the door to confusing, intertwined 'spaghetti logic.' BASIC is a good language for fast, 'one-of-a-kind' development, but is probably not the best choice in this case. FORTRAN, as it's evolved over the years, has much more structure than it did in its early days. The latest versions of FORTRAN allow one to write programs in the modular form of objects, but its principal strength is its ability to facilitate large, number-intensive computations. It permits, but does not enforce, structured programming the way C and C++ do.

"In the final analysis, I'd go with either C or C++. The streamlined instructions available in C++ would make it my choice of the two. Just about any microprocessor we choose will have a C++ cross-assembler available for it, since C++ has by default become the standard object-oriented programming language these days. That's my recommendation."

"Makes sense to me," said Anne. "I was actually thinking along those lines but wanted your confirmation before proceeding. How's your C++?"

"Not bad for a circuits person."

"Good, then you can help me debug the code when I get stuck."

"Deal."

Sally headed back toward her lab. Anne set to work writing a C++ program.

EXERCISES

1. Flowcharts like the one in Fig. 5.1 describe the events and logical decision making that go on inside a computer or microprocessor. Engineers often use flowcharts to describe processes and programs.

   a. Draw the flowchart for a computer program that can tally the voting of a five-person city council. The output should indicate the majority of "aye" or "nay" votes with an appropriate output.

   b. Write this program in the computer language of your choice. Include an input mechanism (e.g., the keyboard) for each of the five city council votes.

2. Draw a flowchart for a computer program that could be used to control the traffic at a busy intersection where two streets cross.

a. Traffic should be allowed to flow over the east/west route unless a car stops at the north or south streets entering the intersection.

b. Traffic should be allowed to flow over the east/west route until three cars are stopped at the north street entering the intersection, but only if no car is stopped at the south entrance. If more than three cars become stopped along the east/west route, it should be open to traffic flow regardless of the number of cars stopped at the north/south streets.

3. Computers communicate with other computers and peripherals using either serial or parallel data links. When a parallel link is used, the connection to the computer consists of one wire for each of the bits in the digital data bus, a common return ground wire, plus additional wires for sending synchronizing signals. The latter are needed so that the receiving device will know when to read each digital word sent by the transmitting device.

When a series link is used, data bits are sent one at a time. In a *synchronous link* system, one wire is used by the transmitting device to send the data bits in sequence, one wire is used for return ground, and a third wire is used to send a synchronizing signal. The latter is used by the receiving device to determine the timing between each data bit. In an *asynchronous link* system, typical of the type used to communicate over telephone modems, long-range networks, and the Internet, only one wire pair is available for signal transmission. Data bit synchronization requires that the sending device and receiving device both operate at the same BAUD (for *bits audio*), or bit timing rate. Such timing is never perfect, however; if left uncorrected, the BAUD timing of the receiving device will drift apart from the BAUD timing of the sending device. To ensure that the bit timing sequences will match, the receiving device resets its timer after each digital word. It knows when the received word ends because the transmitting word appends a *stop bit sequence* to each one. After the stop bit sequence is sent, the transmitting device sets the data line to the value **1** as a prelude to sending its next word. It also adds a *start bit* of value **0** to the beginning of each word so that its arrival will be unambiguous.

a. A particular computer sends data in eight-bit packets, or *bytes*. Determine the content of the two bytes represented by the following data sequence. The start bit, from which you can determine the time interval per bit, is shown in bold. The stop bit sequence consists of two data bits held high:

b. Draw the serial data stream for the byte sequence (**1001 1100**) (**0001 1111**) (**1010 1010**).

4. Indicate the type of data link (parallel, synchronous serial, or asynchronous serial) used by a PC to communicate with each of the following devices:

a. Centronics (parallel PC port) printer interface

b. External 28.8 kb modem

c. MacIntosh printer interface

d. Flatbed scanner

e. Electronic piano (MIDI interface)

5. The *pulse width modulation* system used by Sally's motor controller applies full voltage to the motor while adjusting the *duty cycle*, or time interval, over which full voltage is applied. The current that flows when voltage is applied will be determined by the motor speed and internal resistance. The average power consumed by the motor will be equal to the time average of the voltage-current product. The pulse-width modulated waveform must be produced by Sally's thyristor board in response to the digital data signal sent from Anne's computer module. Write a program in C, MATLAB, or the language of your choice that can produce the required voltage waveform. Your program should accept a binary or decimal number between 0 and 255, then produce an output that is high (logic 1) for an amount of time proportional to the ratio of the input to the reference value 255.

6. Analog-to-digital interfacing is an important part of many computer controlled engineering systems. Although most physical measurement and control involves analog variables, most data collection, information transmission, and data analysis is performed digitally. Analog-to-digital (A/D) and digital-to-analog (D/A) circuits provide the interfaces between analog and digital worlds.

   Digital-to-Analog Conversion. A D/A converter produces a single analog output signal, usually a voltage, from a multibit digital input. One common conversion algorithm produces an analog output proportional to a fixed reference voltage as determined by the equation

$$v_{OUT} = n\, V_{REF}/(2^N - 1)$$

   where N is the number of bits in the digital input word, n is the decimal value of the binary number represented by all the input bits that are set to 1 in the digital input word, and V_{REF} is a reference voltage. When n is equal to $(2^N - 1)$, v_{OUT} is equal to V_{REF}.

   a. Suppose that the input to an eight-bit D/A converter is **0010 1111** with $V_{REF} = 5$ V. Find the resulting value of v_{OUT}.

   b. A ten-bit D/A converter is fed the input word **00 1001 0001** and is given a reference voltage of 5 V. What is the output of the converter?

   c. What is the smallest *increment* of analog output voltage that can be produced by a twelve-bit D/A converter with a reference voltage of 10 V if the algorithm shown above is used?

   d. What is the largest analog output that can be produced by an eight-bit D/A converter if $V_{REF} = 12$ V?

   Analog-to-Digital Conversion. An A/D converter compares its analog input to a fixed reference voltage and then provides a digital output word B given by

$$B = int\; v_{IN}(2^N - 1)/V_{REF}$$

   where the operator *int* means "round to the nearest integer." This encoding operation is called *binary-weighted* encoding. A full-scale binary output (all bits set to **1**) is produced when $v_{IN} = V_{REF}$.

   e. An eight-bit binary weighted A/D converter has a reference voltage of 5 V. Find the analog input corresponding to the binary outputs (**1111 1110**) and (**0001 0000**).

   f. Find the binary output if $v_{IN} = 1.1$ V.

   g. Find the resolution of the converter.

   h. Find the additional voltage that must be added to a 1-V analog input if the digital output is to be incremented by one bit.

7. Boolean algebra is a system of logic used by many computers. In Boolean algebra, variables take on one of two values only: TRUE, or logic **1**, and FALSE, or logic **0**. Boolean operators include AND, OR, and NOT. The AND operator is represented by a dot between variables, *e.g.,* Y = A · B · C means: Y is true if A, B, and C are *all* true. The OR operator is represented by plus signs, *e.g.* Y = A + B + C means: Y is true if one or more of A, B, or C is true. The NOT operator, represented by an overbar, simply reverses the state of the variable, *e.g.,* Y = $\overline{A}$ means Y is true if A is false.

   a. Verify the following equations in Boolean algebra:

$$A \cdot B + \overline{A} \cdot B = B$$
$$(A + B) \cdot (B + C) = B + A \cdot C$$
$$(A + \overline{B}) \cdot (A + \overline{C}) = A + \overline{C + B}$$

b. DeMorgan's theorem states that the Boolean expression $\overline{A \cdot B \cdot C}$ is equivalent to $\overline{A} + \overline{B} + \overline{C}$. Similarly $\overline{A + B + C}$ is equivalent to $\overline{A} \cdot \overline{B} \cdot \overline{C}$. Verify both forms of DeMorgan's laws using various sets of test data.

8. Draw a flowchart that will implement the system described in the following memo:

> To: Xebec Design Team
> From: Harry Vigil, Project Manager
> Subject: Bank Automatic Teller Machine (ATM) Simulator
>
> Our client for this project is in charge of teaching programs for elementary school students in a regional school system. The curriculum followed by these students includes practicing such common tasks as calling on the telephone, going to the supermarket, making change, and taking the bus. Their teacher has requested that we develop a machine that can simulate the functions of an automatic teller machine (ATM) such as one might find at any local bank. The availability of such a system will allow the teacher to train students without tying up an actual bank ATM machine.
>
> Your task is to design and build such a simulator using the components and materials of your choice. The details of the simulator's operation should be fine-tuned after your initial meeting with our customer; however, here is a list of specifications that you can use as a guide in preparing your initial technical plan:
>
> • The simulator should be self-contained with no actual contact to the outside world.
>
> • The simulator should realistically simulate such features as user inquiry, prompting for password, type of transaction, and dollar amounts.
>
> • The simulator should include its own set of entry keys or buttons, its own display device, and its own printer and dispenser for (simulated) money.
>
> • The simulator should be triggered into operation by the insertion of an ATM-type bank card. The decoding and interpretation of information stored on the inserted cards is not a necessary feature of the system. An acceptable solution could instead involve storing passwords (possibly as an updateable list) inside the simulator; this list would be activated by the insertion of any card of the appropriate size.

9. MATLAB is a versatile and comprehensive programming environment particularly suited to solving engineering problems. MATLAB provides commands that allow you to quickly plot and observe variables, organize data, and solve systems of linear equations that can be put into matrix form. Start MATLAB from your PC, MacIntosh, or network provider. You should see a MATLAB header and a double "greater than" command prompt $>>$. Type the following, one line at a time:

```
a = 2
b = 3
```

(The spaces are not really necessary; they are added here for clarity.) These lines establish a and b as variables in memory, to which you have assigned the values 2 and 3. Note that a and b will remain in memory as working variables until you clear them with a clear command. You can, however, change their values at any time with re-declaration statements such as a = 6. If you type just the letters a and b, one line at a time, MATLAB will echo their values. Try it. Now type the following two lines, but add a semicolon at the end of each:

```
a = 4;
b = 8;
```

Notice that adding a semicolon suppresses the echoing of the variable value when it is typed in. Try typing the letters a and b, one line at a time, to see what values are echoed back by MATLAB.

Next type the lines a + b, a*b, and a^2 + b^2, all with and without the trailing semicolon, to see what happens. Also try the line c = a + b. The latter establishes c as an additional working variable in MATLAB's memory.

You should now use MATLAB to perform the calculation z = m*x^2 + p*y^3 for the case m = 2.3, p = 3.1, x = 2, y = 4. You should obtain the answer 207.60.

With the variables m, p, y, and x established in MATLAB memory, you will now create a text file using the text editor on your system. If you are in Windows™, Windows-95™, MacIntosh System 7, or MacIntosh OS™, the main MATLAB command window has a prompt "New M file" which you should access to open another edit window. If you are operating from a Unix system, typing !edit, where edit is the command to open whatever text editor you normally use, will open up an edit window.

Type the following lines, then save your file using the syntax filename.m, where *filename* is any name you choose. Be sure to include the .m suffix:

```
for x = 0:2:10;
z = 3*x + b*y + p
end
```

After you have saved the file, return to the MATLAB command prompt, and type filename (without the .m suffix). MATLAB will recognize your file as an .m batch file and process the commands. The first line instructs MATLAB to perform everything until the end command, first for the value x=0, then for values of x incremented by 2 up to the maximum value of 10. The second line has no trailing semicolon, so MATLAB echoes the value of z each time the loop is traversed with a new value of x. Note that the values of p, y, and b that you declared previously are still retained in memory and are used in the calculation. The previously declared value of x, however, is overwritten by the iterated values x = 0, 2, 4, 6, 8, 10. The values of x and z remaining in memory after the iteration are the most recent values computed for x = 10. Try typing x and z from the command prompt to echo their values as stored in MATLAB memory.

Now edit your .m file to include the following additional lines:

```
z=[];
for x = 0:2:10;
z = [z (3*x + b*y + p)]; %Semicolon optional
```

```
end
plot(x,z)
```

Be sure to leave a space after the z inside the bracket in the third line. Save the file, then run it. Note that everything after the % sign is interpreted as a comment by MATLAB. The first line establishes z as a one-dimensional array variable, initially with no entries. The third line appends another entry of value (3*x + b*y + p) to the array each time the loop is traversed. The last line after the end statement opens a graphical window, then plots z as a function of x. After running this program, type z in the command window. Note that all the computed values of z are stored as a one-dimensional array in MATLAB memory.

a. Use MATLAB to compute and plot the position of the car versus time if full acceleration (800 N) is applied to the wheels from an initial rest position (refer to Chapter 4 for a discussion of velocity and acceleration).

b. Plot the power input to the motor versus time under conditions of part (a) if the motor is 90% efficient.

CHAPTER

6

Mechanical Loading and Testing

Wednesday, 8:00 AM ...

Harry had called a design review meeting. The entire team was present, including senior mentors and junior engineers. Dave was reporting the results of his destructive loading tests on samples of materials to be used on the car's chassis frame. A key design decision had to be made soon—what material would be chosen for the frame's main structural members. Although the meeting was an informal one, Dave had carefully prepared overhead slides and arranged his talk as if it were a formal presentation. He wanted no one to misunderstand or misinterpret his data and had carefully followed the techniques he learned during his seminar on effective communication that he took as an undergraduate.

Dave took his place at the front of the room, then began with a short introduction that explained the purpose of his talk. "Thank you all for coming to my presentation. In this talk I will summarize test results on samples of materials that are candidates for the chassis frame of our electric car. As you all know, we've decided to seriously consider composites—matrices of glass, carbon fiber, and epoxy resin—for the principal structural members of the vehicle. This choice will result in some extra costs, because these materials are more expensive than steel, but I think ultimately we'll have a better-performing, and hence better-selling, vehicle. I've done some initial tests on sample compositions and wish to share them with you." Dave displayed the first of his overhead transparencies:

Loading Tests on L-Type Carbon Composites
Dave Jared
Xebec Research and Development Corporation
Mechanical Design Lab

His next slide summarized the content of the presentation, providing an overview to the audience:

Summary of Presentation
- Description of Composite Materials
- Selection of Test Samples
- Stress-Strain Properties (Non-Destructive Test)
- Maximum Load to Yield Point (Destructive Test)
- Surface Hardness Test

"Let me begin by providing a brief review of composites. These materials were invented in the 1970s by the aircraft industry as possible lightweight alternatives to more expensive metals such as titanium and magnesium. Now they're used in everything from bicycles to sailboat masts. Composites are made from a matrix of carbon or glass threads woven into the desired shape and impregnated with high-tensile-strength epoxy resin." Dave passed out several samples of carbon-fiber composite materials. They were black in color and felt like plastic. "At issue here, besides the obvious decision of whether we should go with composites at all—and that's a cost/benefit question—is exactly what composition to choose if we do. 'Composition,' to remind you, is a measure of the percent weight of carbon fiber to epoxy. The diameter of the fibers is also a factor. Basically, the more fiber in the mixture, the more expensive the material. The trick is to find the optimal composition while considering strength *and* cost." I have here some data on carbon composition, since carbon fiber is the type we're most likely to choose. Dave displayed an overhead which described the various composites he had tested:

PRODUCT	%CARBON	APPROX. COST PER POUND
L-8	8	$ 96
L-10	10	$ 102
L-12	12	$ 113
L-16	16	$ 120
L-20	20	$ 136
L-24	24	$ 141

"Above about a 24% fill rate, this particular composite has too little epoxy resin to hold together well," explained Dave, "and below 8%, too little carbon fiber to retain tensile strength. The costs may seem high, but remember, these things are so lightweight, a little goes a long way."

Dave next described the first of the tests performed on the samples. "The first test I performed consisted of determining the stress/strain relationship for each of the materials in the sample list for a number of different fiber diameters. For the benefit of you non-mechanicals, *stress* is a fancy word for applied force, and *strain* is another term for the amount by which the material stretches (or compresses) in response to the applied force. In a *linear* material, the strain, or stretching, is directly proportional to the applied stress. Double the stress, and you've doubled the strain. If too much strain occurs due to too much applied stress, the material will go into its nonlinear region in which strain is no longer directly proportional to the added stress."

Dave picked up a marker and wrote on the white board. "In real life, we usually average the stress and strain over the cross-sectional dimensions of the material to come up with a force versus distance formula of the form

$$F = -kx$$

where x is the displacement, F the restoring force exerted by the material (hence the negative sign), and k the so-called 'spring constant.' The linear region is one where the parameter k is constant. In calculus terms, the linear region can also be defined by

$$dF/dx = \text{constant}$$

For a non-linear material—one that's exceeded its elastic limit—this relationship becomes

$$dF/dx = -g(x)$$

where $g(x)$ is a nonlinear function of x.

"Now let me explain the tests I've performed. The first involves measuring the stress/strain, or force/displacement, properties of the material using the following setup." Dave put up the overhead shown in Fig. 6.1.

"The sample to be tested is first machined into a bar of uniform standard cross-section—in this case, a round cross-section 1/2 inch in diameter. It's then installed in a tensile test machine that can apply a stretching force to the bar. The machine can even apply enough force to physically break the test sample if we wish to do so. I'll be reporting on such tensile failure tests in a bit. For these stress/strain tests, we measure the applied force with something called a 'load cell.' A load cell produces a voltage signal that is proportional to the stretching force applied to the test sample. We measure the strain with something called a strain gage. A strain gage is a device— a thin film resistor, actually—that is glued onto the side of the test sample. It then produces a voltage signal proportional to how much the sample's been stretched from its initial length. In our setup, we send the load-cell and strain gage signals to a data recorder, where they're stored for future display, tabulation, or plotting.

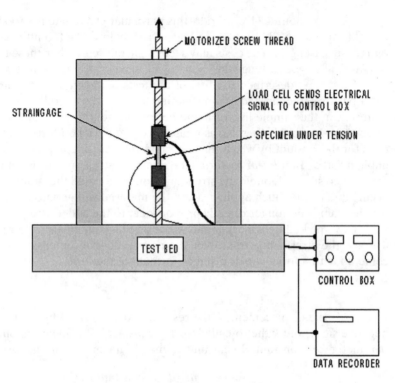

Figure 6.1 Dave's tensile test machine and computerized data recorder

"To run a test, we begin with zero force, then slowly ramp up to the largest force we wish to apply, periodically recording both force and displacement—stress and strain—along the way. I've written a program in LabView™, a popular instrument control and data management software package, to display test results in graphical form on the PC screen or on paper printout." Dave put up the slide shown in Fig. 6.2. "Here's a typical plot of data taken on one sample, in this case, L-8:

"Note that the slope is measured in kilo-pounds per mil of displacement, or elongation. Remember that a mil is 0.001 inch, so it's a very small distance. In the linear region, the slope of the curve is equal to the elastic constant k. For our application, we'd like as high a force per unit displacement value as possible, so that the frame will be nice and stiff. We will, however, need a yield stress of at least 500 MPa (mega-Pascals) to ensure an adequate safety margin of about five times the largest force we realistically expect on the frame. The yield stress defines the point at which the material is no longer able to hold back against the applied tension."

A Pascal, wondered Keith. Then he remembered. *Oh, yeah. That's an SI unit of force per unit area, or pressure, measured in Newtons per square meter.*

Dave continued. "A *test matrix* consists of a table of methodically performed tests in which one variable changes on the vertical axis and one along the horizontal axis. Let me show you the results of measurements obtained from one of my test ma-

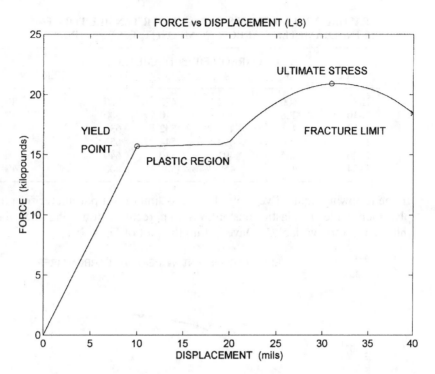

Figure 6.2 Typical force/displacement curve

trices." Dave put up the following table on the screen. He'd copied it from his engineer's logbook, where it served as a written record of the data:

**MEASURED ELASTIC CONSTANTS (kN/mm) UNDER TENSILE FORCE
0.5 IN. DIAMETER CARBON COMPOSITE TEST SAMPLES**

| | CARBON FIBER DIAMETER (mils) | | | | |
SAMPLE	3	4	5	6	PERCENT CARBON
L-8	280	236	216	200	8%
L-10	299	253	220	206	10%
L-12	316	273	230	216	12%
L-16	323	293	240	226	16%
L-20	330	306	250	236	20%
L-24	336	323	260	246	24%

"The entries in this first test matrix table indicate the elastic constant, m~~~~ured in units of kilo-Newtons per millimeter, for each sample. Note that I~~ converted from English units of pounds and mils to metric units of Newto~ ~d millimeters. My next table gives the yield stress of each sample in mega-P~ ~als. The 'Pascal,' defined as Newtons per square meter, is a unit of pressure. ~ ~orce per unit area.

MEASURED YIELD STRESS (MPa) UNDER TENSILE FORCE
0.5 IN. DIAMETER CARBON COMPOSITE TEST SAMPLES

SAMPLE	CARBON FIBER DIAMETER (mils)				PERCENT CARBON
	3	4	5	6	
L-8	400	432	472	560	8%
L-10	428	460	510	600	10%
L-12	445	480	532	631	12%
L-16	460	499	544	651	16%
L-20	455	491	540	640	20%
L-24	428	459	506	600	24%

In the following figure, I've plotted the most important parameter from the force/displacement table, the elastic constant, versus percent carbon using fiber diameter as the third parametric variable." Dave put up the plot of Fig. 6.3:

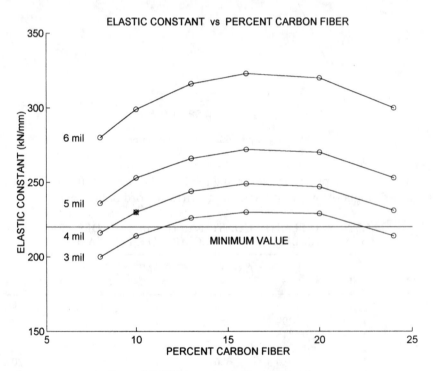

Figure 6.3 Elastic constant versus percent carbon

"In my next overhead, I've plotted the yield stress versus percent carbon, again using fiber diameter as the parametric variable." Dave changed his slide to the one in Fig. 6.4. "Because we need a yield stress of at least 500 MPa, you can see that we're limited to fiber diameters of 5 mils or more. We'd like a large force/displacement ratio (a small elongation per unit force) to make the frame stiff—at least 220 kN/mm. As you can see from the previous slide, the largest force/displacement ratios occur for 6 mil fiber. The

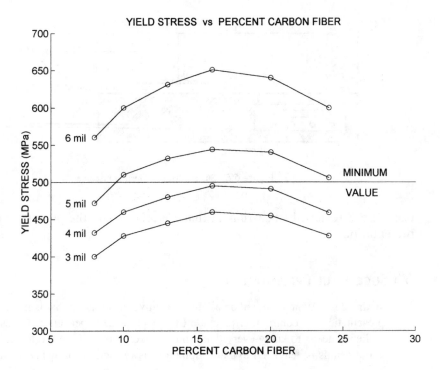

Figure 6.4 Yield stress versus percent carbon

5 mil fiber, however, has adequate properties, and we can save some money by going with that choice. The largest force displacement ratio occurs with 16% carbon, but even 10% carbon will give us an elastic constant above the minimum needed value. So, in summary, I'm recommending that we go with 10% carbon composite with 5 mil fiber, at a present cost of about $102 per pound.

"One last test. It's important that we have a feeling for the hardness of the material, that is, how resistant it will be to scratches and dents. This property has nothing to do with strength—for example, fragile glass is much harder than robust steel—but the harder the material, the more resistant it will be to wear at joints and mating surfaces where friction occurs. I was able to measure hardness by using a machine that dents the samples with a diamond-tipped stylus and measures the force required to achieve a cut of specified depth. Here's a picture of the machine that I used." Dave put up yet another slide, this time the one shown in Fig. 6.5.

All of our samples have about the same hardness, because this property is largely determined by the epoxy which is common to all. I'm measuring hardness values on the order of R-120, using the 'Rockwell' scale, for all the samples. Well, that's about it. Thanks for your attention. Any questions?"

Dave's presentation had been excellent. Harry, his manager, seemed pleased, and the rest of the attendees appeared satisfied with Dave's conclusion. His well-organized slides, carefully presented data, and well-articulated explanations have done the job. The steps that went into Dave's successful presentation are easy to

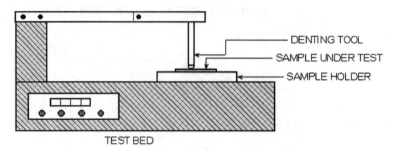

Figure 6.5 Surface hardness testing machine

learn. Let's review them, assuming that you, too, are in the process of preparing a presentation:

STEPS TO A SUCCESSFUL PRESENTATION

Keep it simple.   Avoid cluttered detail and repetitive information that is not essential to the presentation. A common error made by inexperienced speakers is to assume that the audience needs to know every detail the speaker knows about the topic. Clarification about details is better left to a question and answer period following the talk.

Identify the type of presentation.   In the setting of an engineering company, two types of presentations are most often found. One is an informal, unrehearsed presentation intended as a working session for technical groups, project teams, co-workers, and focus groups. The other is a formal, prepared presentation appropriate for external audiences, upper management, customers, or clients. Dave's presentation, prepared to address Harry and the senior mentors, was of the latter category.

Know your audience.   A talk should be planned for an appropriate level of detail. If the audience has technical expertise that's similar to yours, it's appropriate to include information about tests, materials, background theory, and design calculations. If the audience has minimal technical background, you should not dwell on these topics as they will be lost to the listener and will lead to a tiresome talk.

Have good information to present.   A presentation should always convey a clear message. The talk should include valuable information for the listener. Dave did a careful job of deciding which tests were necessary to answer his design questions. He constructed a *test matrix,* verifying key parameters sequentially over approximate ranges. He systematically stored the data and made accompanying entries into his engineer's logbook.

Prepare well-laid-out slides and overheads.   In general, the simpler the slide, the greater its impact. Avoid the temptation to put every available bit of information on overheads and other visual aides. A slide should never include so much information that it becomes crowded and hard to read. Slides and overheads should be highlighted with "bullets" or other appropriate icons.

Include all parts of the talk. Like a well-written essay, a good talk consists of an introduction, body, summary, and conclusion. The purpose of the talk should be clearly stated in the introduction. The most common error made by poor speakers is to assume the audience knows the purpose of the talk. Dave's talk, for example, began with, "Thank you all for coming to my presentation. In this talk I will summarize test results on samples of materials that are candidates for the chassis frame of our electric car."

Although most of the listeners probably knew this information already, Dave took no chances—he recited the obvious for the benefit of any uninformed listeners. It's always a good idea to review the purpose of the talk, because even well-informed listeners like to have their assumptions confirmed.

Summarize the information to be presented. Begin a talk by summarizing the key points to follow. Show your outline in the beginning. The first visual aid or verbal comments should provide an overview of the rest of the talk. This important technique helps the listeners to follow the logical flow of your presentation. You should also give an overview of what the presentation will address. Assume the listener knows nothing. Most will know less about the topic than you think.

A single anecdote at the beginning also may be helpful. Sometimes a short story or humorous remark sets the audience (and the speaker) at ease.

Try not to read your talk. Although notes are sometimes necessary, never read them. Maintain eye contact with your audience. One good technique is to use your own slides or overheads as visual cues instead of written notes. Notations, out of sight of the viewer, can be written in the margins of the overheads.

Ask your own questions, then answer them. This technique is useful for emphasizing key points in the talk. Highlight questions on visual aides with "bullets" (•) where appropriate.

Do not use equations. This rule can be broken occasionally. Most listeners will not have the time to read and digest equations. If an equation is well known, for example, Newton's law $F = ma$ or Ohm's law $V = IR$, it may be appropriate to include it to set the context of the presentation. Most often, simply citing the relevent equation by name is sufficient. Examples include Bernoulli's equation, the Pythagorean theorem, Brunell hardness formula, Coulomb's force law, Laplace's equation, the Navier-Stokes equation, etc.

Respond to audience questions. Restate questions so that the entire audience can hear them. Restating a question will also help to clarify its content and will give you an extra moment to formulate your response.

Be straightforward. Be willing to admit to not having previously thought about something. "I don't know," or "I'll get back to you," are appropriate responses in lieu of inappropriate handwaving, inaccurate answers, or flimsy explanations.

Keep it short.   Plan to use between 50% to 60% of the allotted time. Most presentations are too long. A good speaker presents just enough information to convey the key points.

Let the audience know when the talk is over.   Avoid the awkwardness of a hanging ending by clearly identifying the end of your talk. Dave ended his presentation, for example, with a simple, "Thank you. Any questions?" A slide that says "thank you" or "the end" is also acceptable.

    The closing of a talk is often preceded by a conclusion which summarizes its key points. The conclusion often takes the form of a recommendation or a sales pitch. Dave's conclusion, for example, included a recommendation on which material to use: "So, in summary, I'm recommending that we go with 10% carbon composite with 5 mil fiber, at a present cost of about $102 per pound."

    Note that Dave's conclusion was not the very last sentence of his presentation. It's acceptable to place the conclusion near, but not at, the end if it helps with the flow of words. In Dave's case, his parenthetical remarks about the hardness test provided a nice, informal ending to his presentation.

Keep it simple.   Remember, simplicity is the key to successful speaking.

REFERENCES

1. LabView™ is a registered trademark of National Instruments, Inc.

EXERCISES

1. Cite the features of Dave's presentation that correspond to each of the "Steps to a Successful Presentation" enumerated above.

2. About a month after Dave's presentation, Harry scheduled a dry run of a progress report to be presented to the project investors. Rewrite Dave's load test presentation so that it is appropriate for a presentation to the investors. Incorporate into your revision the additional constraints of this unique type of presentation.

3. Pretend that you are Thomas Edison. Prepare layouts for visual overheads that you would use to describe the invention of the electric light bulb. For the purpose of this exercise, which focuses on presentation techniques, you may fabricate the data and experiments that Edison was likely to have recorded.

4. Data presentation is an important part of any talk. Shown below are measurements on the displacement of a test sample versus force during compressive loading tests. Plot this data on a graph using linear scales for both the horizontal and vertical axes. You may plot by hand on graph paper or use an appropriate computer software package such as MATLAB, Excel, VisiCalc, Lotus, etc.

Force (kN):	1.0	2.0	3.0	4.0	5.0	6.0	7.0	8.0	9.0
Displacement (mm):	0.20	0.40	0.60	0.80	0.88	0.95	1.00	1.05	1.09

5. If an initially charged capacitor is suddenly connected to a resistor, its voltage will decay exponentially. The discharge can be represented by the equation $v(t) = V_o e^{-t/RC}$, where V_o is the initial

voltage, C the capacitance, and R the resistance. Shown below are data taken on a 100 μF capacitor initially charged to 16 V and connected to a 1 kΩ resistor.

Time (ms):	0	20	40	60	80	100	120	140	160	180
Voltage (V):	16	13.1	10.7	8.8	7.2	5.9	4.8	3.9	3.2	2.6

a. Plot the data as a function of time and verify that the decay looks exponential.
b. At what time will the voltage reach zero? At what time will it reach 1% of its initial value?
c. Now plot the capacitor data using a semilog scale for the vertical axis. Note that if we take the natural log of both sides of the capacitor equation, it becomes a straight-line equation as a function of time:

$$\ln v(t) = \ln V_o e^{-t/RC}$$
$$\text{or } \ln v(t) = (\ln V_o) - t/RC$$

Show that your plot of $\ln v(t)$ versus time yields a straight line.

6. When an initially *uncharged* capacitor is suddenly connected to a resistor and battery in series, the capacitor voltage will *rise* exponentially. The discharge can be represented by the equation $v(t) = V_o(1 - e^{-t/RC})$, where V_o is the battery voltage, C the capacitance, and R the resistance.
a. Compute sufficient points to plot v(t) as a function of time for the case $V_o = 24$ V, $R = 100$ kΩ, and $C = 2.2$ μF. Use linear scales for each axis.
b. Now plot your data using a logarithmic scale for the vertical axis. Show that your plot yields a straight line.

7. A spring is measured on a tensile test machine. From the following data set, determine the elastic constant of the spring in its linear region. In other words, find the value of k in the equation $F = -kx$.

Force (N):	10	20	30	40	50	60	70	80	90	100
Displacement (mm):	2.2	4.4	6.6	8.8	9.7	10.4	11.0	11.6	12.0	12.3

8. Test data on a compressed air tank is shown below. The tank is heated by placing it in a hot oil bath and the pressure is measured with an external gauge. Plot the given data, verify the ideal gas formula $PV = nkT$, and determine the value of the product nk, where k is Boltzmann's constant. Note that °F is equal to 1.8(K − 241), where K is the temperature in Kelvin.

Temperature (°F):	57.6	70.2	106.2	142.2	178.2	214.2
Pressure (PSI):	20.0	20.5	22.0	23.4	24.9	26.4

9. Engineers work in all sorts of units, including SI (kg, Newton, meter, second), English (slug, pound, foot, second), and metric (gram, Newton, centimeter, second). Convert each of the following quantities to its SI counterpart:
a. length: 144 inches to m
b. area: 230 square miles to m^2
c. volume: 263.8 gallons to m^3
d. velocity: 55 mph to m/s
e. velocity: 88 ft/s to m/s
f. force: 312 slug-foot/sec^2 to kg-m/s^2
g. force: 100 lb to Newtons

    h. pressure: 14.5 psi to Pa (N/m^2)
    i. energy: 300 ft-lb to N-m

10. Perform each of the following computations, expressing the answer in appropriate SI units:
    a. (102 mm)(31 kN)
    b. $(1500 \text{ kg})(10 \text{ mm/s}^2)$
    c. $(13 \text{ gm})(5.0 \text{ mm})/(10 \text{ ms}^2)$
    d. $(1.2 \text{ GPa})(87 \text{ cm}^2)$
    e. (55 mph)(2 hr)
    f. $(18 \text{ lb})(0.6 \text{ ft/s}^2)$

11. A rectangular bar having cross-sectional dimensions 10 mm × 40 mm supports a tensile load. If the stress in the bar cannot exceed 120 MPa, what is the maximum allowable load in Newtons?

12. Suppose that a hollow circular cylinder with a wall thickness of 1 inch must support a total axial compressive load of 1300 pounds. The maximum allowed compressive stress on the material is 11.7 kpsi. What should be the minimum outside diameter of the cylinder?

13. Tensile test data for a particular specimen is shown below. Plot the elongation versus load and find the yield limit of the material.

LOAD (LBF)	ELONGATION (MILS.)
500	0.1
1,000	0.3
3,000	0.95
5,000	1.7
6,000	2.0
6,500	2.3
6,700	2.4
6,800	2.7
6,900	3.2
7,000	4.5
7,200	5.9
7,600	8.4
8,400	13.2
9,200	19.0
10,000	25.3
11,200	55.4
12,700	fracture

14. A 100 mm rod of material has reference marks spaced 20 mm apart. The specimen is placed in a tensile test machine and subjected to tension until fracture. After the test, the reference marks are located 24.1 mm apart. What was the percent elongation of the rod just before fracture?

15. The linear relation between force and displacement is often expressed in the distributed form of *Hook's Law:* $\sigma = E\epsilon$, where σ is the stress, or force per unit area, E is the elastic constant, or *modulus of elasticity,* and ϵ is the strain expressed as an elongation per unit length. If a piece of material of length L, for example, is stretched by an amount δ when stress is applied, the strain will be δ/L. Note that strain is a dimensionless quantity.
    a. Suppose that a rectangular bar of length 1.4 m and cross-section 10 mm × 20 mm is subjected to an axial tensile force of 100 kN. If the length of the bar under stress increases by 1.2 mm, find the strain.

b. A rod of circular diameter 30 mm and length 1.5 m is subjected to an axial tension of 85 kN. If its modulus of elasticity is 70 GPa, calculate the total elongation δ of the bar.

16. Engineers who work with forces invariably use vector analysis to compute the net force acting on a body. The end of a boom is subjected to the force of three separate cables. Force F_A exerts a horizontal force of 400 N, F_B an upward force of 250 N inclined 45° from the vertical, and F_C an upward force at 37.8° from the vertical and 82.8° from F_B. Draw a vector diagram of the forces involved. What is the magnitude and direction of the net force acting on the end of the boom?

17. Dave prepared the plots of Figs. 6.3 and 6.4 using the MATLAB software package. A printout of his program used to produce Fig. 6.3 is shown below.
   a. Rewrite the program so that it produces the plot of Fig. 6.4.
   b. Can you write a program in MATLAB that produces the plot of Fig. 6.2?

```
%PROGRAM TO PLOT FORCE/DISPLACEMENT RATIO vs ELONGATION
%fiber diameter (mils):
%   3  4  5  6
L8 = [280 236 216 200];
L10 = [299 253 230 214];
L13 = [316 266 244 226];
L16 = [323 272 249 230];
L20 = [320 270 247 229];
L24 = [300 253 231 214];
PC = [8 10 13 16 20 24]; %percent carbon
hold off;clg
axis([5 25 150 350]);hold on
%Plot elastic constant vs percent carbon
for n=1:4;
y=[L8(n) L10(n) L13(n) L16(n) L20(n) L24(n)];
plot(PC,y); hold on
plot(PC,y,'o');
end
text(6,280,'3 mil')
text(6,236,'4 mil')
text(6,216,'5 mil')
text(6,200,'6 mil')
title('ELASTIC CONSTANT vs PERCENT CARBON FIBER')
ylabel('ELASTIC CONSTANT (kN/mm)')
xlabel('PERCENT CARBON FIBER')
```

7

Breadboarding and Testing

Tuesday . . .

It had been three weeks since the initial kickoff meeting for the electric car project. After numerous conversations with Anne about the software and overall control functions of the car, Sally had finished a first cut at the main circuitry of her electric power controller. The power controller was a circuit made from electronic devices called *thyristors*. It would control the flow of electricity between the battery bank and main drive motor. Without the power controller, the batteries would have to be connected to the main drive motor by simple switches, permitting motor operation at two power levels only: zero power and full power. Sally sat back and thought about the consequences of such a primitive power control system. She made the analogy to a gasoline powered automobile whose accelerator pedal could only be set to zero or full throttle. What an interesting ride that would be!

Sally's electric power controller would also become part of the car's brake system. Following up on an idea that had germinated during Harry's initial brainstorming session, the design team had committed itself to a system whereby kinetic energy stored in the moving car would be converted back to electrical energy and returned to the battery during braking. In the brake mode, the motors would be run as generators. After obtaining a signal from Anne's computer module that the brake pedal was depressed and deceleration was detected by an imbedded sensor, the power controller would set the direction of power from the motor to the batteries. Sally reviewed the photocopy of Anne's logbook page, shown here as Fig. 7.1, which she had pasted into her own logbook. Carefully, she reviewed the diagram again and thought about each mode of the power controller.

For over a week, Sally had been testing her electrical design concept by simulating the principal elements of the power control circuit on a software tool called SPICE.

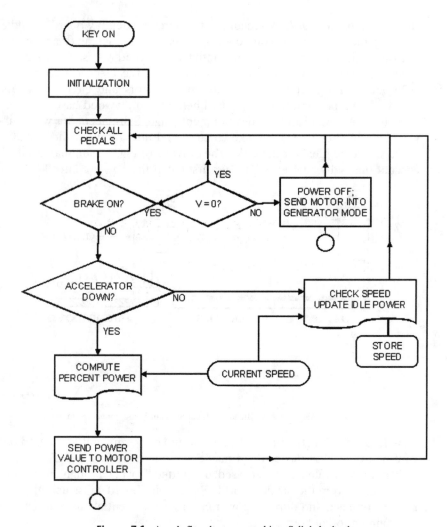

Figure 7.1 Anne's flowchart as pasted into Sally's logbook

Xebec had chosen this accepted industry standard as its circuit simulator, and it was available from a UNIX server via a network link to Sally's desktop workstation. After many paper iterations and simulations on SPICE, she was now ready to build and test an actual version of the circuit using real electronic components. It was common practice at Xebec, as in many companies, to develop electronic circuits by first doing only rough hand calculations and then fine tuning the design by simulating the circuit in software. There was no substitute, however, for actually building the circuit to make sure that it worked.

Sally truly enjoyed building circuits. To her, simulation was a fun computer exercise but seemed a bit phony compared to building and testing with real parts. As the morning progressed, she found herself gathering the necessary materials to build the circuit on a *breadboard*. A breadboard, or circuit assembly board, would allow her to wire together the various components of her circuit, such as resistors, capacitors, and integrated circuits, by plugging them into holes aligned with spring-loaded clips located inside the breadboard. The clips also provided the electrical connections between components. Larger devices, such as the thyristors and circuit breakers, would be mounted on a perforated epoxy board held over a plywood base by ceramic standoff insulators. At the prototype development stage, Sally needed a wiring medium that would enable her to easily make changes and alterations to the circuit. The breadboard would be ideal for the task. The drawing of Fig. 7.2 summarizes the breadboard concept that Sally envisioned for her first cut at the power controller.

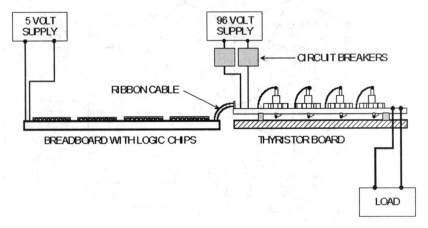

Figure 7.2 Breadboard and thyristor board connected to load

In production, the finished circuit would be permanently soldered onto a printed wire board such as those found inside computers, radios, and TVs. The entire controller circuit would also be encased, or "potted," inside a tough epoxy compound that would seal it from the outside world. Potted wire board construction would produce a robust, compact, and durable power controller—essential characteristics in the harsh environment of an automobile.

With the assistance of her technician who prepared the plywood mounting board, Sally wired the control portion of her circuit and connected all the components, using her own hand drawn schematic diagram as a guide. She then wired up the thyristor board—there were four thyristors in all—and connected it to the control circuit. Lastly, she connected the thyristor circuit to a benchtop power supply that functioned as the car's 96-V battery bank. She carefully checked and tightened these latter connections, because they would be carrying all the current consumed by the load. A low-valued resistor of high wattage rating served as a temporary electrical load in lieu of a drive motor. Sally also connected a 5-V power source to the logic circuit on the breadboard. Her integrated circuit logic devices, typical of most, were designed to operate from a 5-V dc power source. In the final version of the controller, Sally would need to design a separate circuit to obtain the required 5 volts directly from one of the 12-V batteries in the 96-V battery bank.

When she was finished with the wiring, Sally connected diagnostic equipment: an oscilloscope in parallel with the load to measure the voltage delivered versus time, an ammeter in series with the load to measure the current, and a multichannel logic analyzer to monitor the digital signals inside the control circuit. The breadboarding and connection task took the greater part of the morning. Her complete setup, including monitoring instruments, looked like the one shown in Fig. 7.3.

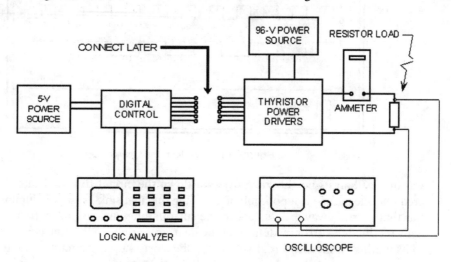

Figure 7.3 Complete test setup with instrumentation

To start off the testing, Sally temporarily disconnected the logic circuit, isolating it from the thyristor board. Disconnecting the thyristor board was a good precaution. Sally did not want to burn out the expensive thyristors should the logic circuit be miswired or malfunction and send the thyristors false firing signals. A fundamental design error by Sally could also result in faulty signals being sent to the thyristors. Sally powered up the logic circuit with the 5-V supply and examined several screens of data from the logic analyzer. A logic analyzer is a device that displays plots versus time of the various digital voltages inside the logic circuit. A sample of the logic analyzer screen observed by Sally is shown in Fig. 7.4. Most of the circuit checked out, but Sally did find a wire that she had forgotten to connect between two of the integrated circuits. After adding the wire, and correcting one other crossed wire, the circuit appeared to function as she had intended.

Satisfied with the operation of the logic circuit, Sally next connected the ribbon cable carrying signal lines to the thyristor board and adjusted the control circuit to deliver a small amount of power to the load resistor. It was important to bring up the power gradually. If anything were wrong with the thyristor board, Sally would be able to detect it without burning out a thyristor due to excessive current and heat generation. Detecting no obvious anomalies, Sally increased the power level, causing the thyristor board to deliver about one quarter of its maximum available power. She left the circuit in this mode for about an hour to let it settle down and come to an equilibrium temperature. When she returned, she indeed noticed that one of the thyristors

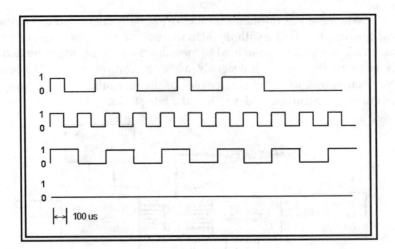

Figure 7.4 Sample logic analyzer screen

was hotter than its neighbors. Sally shut down the circuit to investigate. She discovered that she had done a poor job of attaching the heat sink—a set of aluminum cooling fins that carries away excess heat to the surrounding air—to that particular thyristor. Sally rectified the error by detaching the heat sink, applying thermal conduction grease to its mating surface, and reattaching it. She started up the circuit and left it running for another hour. When she returned, all the thyristors operated at a uniformly modest temperature. She double checked their temperatures with a thermal probe. Satisfied with the first power level, Sally increased it by about twenty percent, then left the circuit running for another hour. She continued in this way, slowing turning up the power, checking and confirming its operation at each level. By the end of the day, she had the unit supplying full power continuously to the resistor load.
The next day...

Sally had arranged to meet Dave in the mechanical test lab. It was time to connect the power controller to a real motor. Their test bed, prepared by Dave, is shown in Fig. 7.5. It included a bank of eight 12-V lead-acid car batteries, for a total of 96 V,

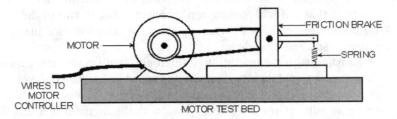

Figure 7.5 Dave and Sally's motor test bed

a drive motor, and a mechanical loading apparatus called a friction brake. Although these batteries were not the ones that would go into the actual vehicle—the finished vehicle would employ state-of-the-art battery technology-they were of a type that was easy to find at the local auto supply store and would suffice for now. Earlier in the week, a technician had built a support rack for the batteries and brought heavy cables (#0 wire, per Sally's specifications) out to a power terminal bus. The motor was about the size of two large coffee cans stacked end-to-end and had a 3/4 inch diameter shaft. The friction brake was capable of applying a frictional load to the motor shaft to simulate the force exerted on the shaft by the car wheels. The entire setup reminded Sally of the frictional loading device on her stationary exercise bike.

Sally's digital logic board was still connected to a separate 5-V power supply that she had brought from the electrical test lab. Eventually, she would need to modify the power controller board by adding a circuit capable of producing the needed 5 V from one of the 12-volt batteries. As she had done with the resistive loading tests, Sally used an oscilloscope, ammeter, and logic analyzer to monitor the various voltage signals inside the power controller as well as the voltage going to the motor. She and Dave checked the connections, then started testing the system.

At about the half power point, Sally noticed an anomaly on the oscilloscope trace. The waveform delivered to the motor was supposed to be a pulse-width modulated waveform—a rectangular-shaped voltage pulse of constant height but of varying time duration. The actual waveform included some small wavy ripples that occurred whenever the voltage jumped from its low to its high value. As shown in Fig. 7.6, these

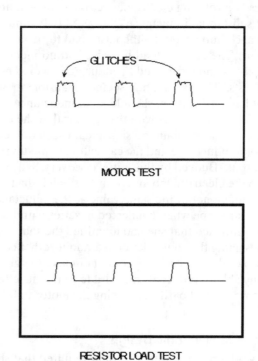

Figure 7.6 Oscilloscope waveforms from resistor and motor load tests

wavy lines, or "glitches," occurred every time one of the thyristors switched on. Sally had not observed any glitches during either the resistor load tests or the SPICE simulations. She compared the present waveform to the same waveform measured during the resistor load test. She had saved a picture of the latter using the disk drive of the digital sampling oscilloscope. Indeed, the glitch she now observed seemed to be related to the presence of the motor and its load.

Sally was puzzled. She checked and rechecked everything: wiring, connections, load conditions. She knew that the glitch could turn into bigger problems later on. The spurious waveform might lead to unwanted current transients in the thyristors, causing them to overheat and fail prematurely. Sally disconnected the logic control board from the thyristors and tested the logic board separately. It functioned normally. She rechecked the connections on the thyristor board and motor load. Nothing seemed out of place, and all the connections were intact.

Sally was stumped. She decided to consult her senior mentor about the problem. "I'm picking up a glitch on the motor waveform that did not show up on either the SPICE simulations or the resistor load tests," she explained. "Frankly, I've run out of ideas. Do you have any?"

"Have you thought about the effects of motor inductance on the thyristor circuit?" offered the senior engineer. Inductance is created whenever a wire is formed into the shape of a coil, as it was inside the motor. The rotating armature of the motor was basically a rotating electromagnet made by winding wire around an iron core. It surely had substantial inductance. Sally's SPICE simulation had not taken internal motor inductance into account—the latter had to be included in SPICE as a parasitic circuit element. The resistor load likewise had no inductance of significance, so that the glitch, if it were indeed caused by inductance, would not have shown up during the resistor load tests. The explanation seemed plausible to Sally.

Back at her workstation, she added an inductance element to her SPICE circuit description, placing it in series with the resistor that she had used to represent the motor. The "glitch," really a damped sinusoidal oscillation caused by the interaction between the motor inductance and the capacitors of the thyristor circuit, appeared on the simulation output. Buoyed by this initial discovery, Sally brought her controller breadboard back to the electrical lab and found a small inductor lying around in the surplus parts bin. Although not of the same value as the inductance of the motor, it produced a glitch of similar shape when connected in series with the resistor load.

Sally was satisfied that she had identified the source of the problem. She spent a few days redesigning the controller board, adding other components to compensate for the inductance of the motors. She could not make the glitch disappear completely but was eventually able to reduce its magnitude so that its effects would be negligible. By the following week, the board was driving the motor to full power. The glitch problem was solved.

The Tools of Electronic Design

Sally's work on the power controller required that she employ many of the techniques and procedures commonly used in electronic design today. The following sections describe the work environment for electronics development in more detail.

Breadboarding and Prototype Construction

Breadboarding is used whenever a circuit is developed for the first time. For rapid prototype development in which many design changes will be made, component breadboards are ideal. The breadboard of Fig. 7.7 is typical of most versions. Holes along

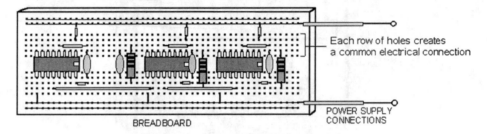

Figure 7.7 Breadboard and circuit connections

each short row in the board reside over conducting metal spring clips into which component leads can be inserted. All leads inserted into the same row will be electrically connected. Additional connections across rows can be made using jumper wires. Most breadboards also contain long outer rows of interconnected holes that are used as the power supply distribution connections. Holes on the typical breadboard are spaced at 0.10 inch, which is the same spacing as the pins on most integrated-circuit packages. The leads of discrete components such as resistors, capacitors, and transistors can be cut to size and bent at right angles, so that the components are easily inserted.

Permanent Wiring Methods

When a breadboard circuit has been tested, debugged, and otherwise perfected, it's ready to be produced in permanent form. Virtually every commercial, mass-produced circuit is wired together using an etched, copper-clad *printed-circuit board* (sometimes called a *printed wire board*). As illustrated in Fig. 7.8, the wiring paths on a "PC" board (abbreviation not to be confused with "personal computer") consist of thin strips of copper, called *traces*, bonded to an insulated board made from phenolic or epoxy glass of approximately 1 mm thickness. In conventional PC board design, component leads pass through holes drilled in the board and are soldered to the copper wiring paths. Technology exists to make PC boards with several separated layers of copper, so that wiring paths may cross without making contact. More modern boards make use of *surface mount* technology in which the "leads" of components consist of short, stubby connection pads that solder directly to the copper PC board traces.

Printed-circuit board methods are ideal for mass-producing circuits. Many commercial software packages exist to aid the designer in laying out the board for subsequent etching by a commercial PC board vendor. These software programs generate files that can be fed directly to a board fabrication machine. The per-unit cost of a mass-produced, professionally made PC board is low, but the setup charge is high because sophisticated fabrication equipment is required. Hence PC board methods are generally used only for circuits that will be produced in large numbers. One alternative for prototyping involves a system in which PC boards are produced by hand in

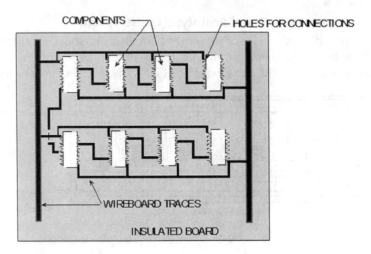

Figure 7.8 Printed-circuit board with components connected

"one-at-a-time" fashion. The paths of copper traces are first marked with a resist pen or marking tape on a copper-clad board. The board is then etched by hand in a liquid etchant that dissolves unmarked copper. After etching, the traces are tinned (covered with solder film so that they are more easily soldered) and holes are drilled for component leads. Relatively inexpensive kits for hand etching boards are available from most electronics parts vendors. Hand etched boards provide the physical robustness of a mass-produced PC board, but are of significantly poorer quality, require traces of much larger line width, and require much more time to make. Correcting errors or making changes in the circuit layout is also difficult.

A form of PC board called copper trace board, or sometimes experimenter's prototype board, is also commercially available. This product is similar to a PC board and contains pre-etched, predrilled strip traces to which component leads may be soldered. The layout of predrilled holes in an experimenter's board resembles that of a similarly sized breadboard, making the transition from temporary to permanent wiring easy. Connections across traces are made by soldering wire jumpers into place. Changes to the circuit can be made, but a given connection can be resoldered only a few times before the trace begins to separate from the board.

One-of-a-kind prototypes may also be made using wirewrap technology. As depicted in Fig. 7.9, components are inserted into special sockets that have long pins pro-

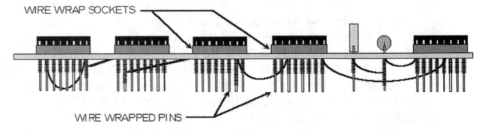

Figure 7.9 Wire wrap board

truding through holes in a prepunched insulated board. Single wirewrap pins are also available for mounting discrete components such as individual resistors and capacitors. Connections between pins are made by tightly winding special wire around the pins with a wirewrap tool. The pins used for wirewrap have a square cross section that digs into the wrapping wire and makes a good metallurgical bond. Wirewrapped circuits are not as robust nor nearly as compact as PC-board wired circuits, but still provide good, nearly permanent connections. Their principal advantage is the relative ease with which connections can be altered to accommodate design changes and alterations.

The Parts Inventory

When a design is ready for mass production, a list is created of all parts, no matter how small or minor, that are used in the circuit and its package. This list is used by many people, including those responsible for ordering or making the parts, determining overall cost, and assembling the finished product. During the initial or prototype development stages, however, it is difficult to work from a fixed parts list. Even minor changes in the design may require components to be changed or altered. Waiting for new parts to be ordered on an as-needed basis can significantly delay the design process. A well-equipped laboratory will include a running inventory of commonly used parts. Maintaining a parts inventory may involve added expense, but the latter is offset by a considerable savings in time.

Avoiding the "Bird's Nest" Pitfall

Good wiring practice requires that a circuit be compact, neat, and orderly, with all leads cut as short as possible. Component bodies should physically rest on or just above the board surface, and wires should be easy to trace and touch with the probe of an oscilloscope or multimeter. The "bird's nest" approach illustrated in Fig. 7.10 should be avoided at all costs. When a circuit consists of a disorderly tangle of wires leading haphazardly in every direction, component leads may short together, wiring errors are likely, and circuit testing becomes extremely difficult. One easily becomes lost in such a circuit. Long leads also create unnecessary stray capacitances and inductances which may adversely affect circuit behavior. A sloppy circuit also affects the attitude of the designer, who is likely to take the design process less seriously if work on the circuit is difficult. The wise designer produces circuits that are neat, compact, tidy, and easily accessible.

Documentation

When designing an electronic circuit, the engineer must keep careful records of all tests performed and design elements completed. As in other engineering design disciplines, it is a good strategy to write everything down in a logbook even if it may seem unimportant at the time. Documentation should be written in such a way that another engineer who is only slightly familiar with the project can reconstruct the circuit and repeat all work done by simply reading the logbook. Careful documentation will aid in the task of writing product literature and technical manuals should the design be destined for commercial sale. Above all, good documentation will provide the engineer with an overview of the design history and the key questions that were addressed during the design process.

Figure 7.10   The Bird's Nest

ELECTRONIC DESIGN PROBLEMS

Design problems included in the following "Exercises" section may be developed on paper or may be built using electronic components readily available at local or mail-order electronic supply stores.

EXERCISES

1. A circuit breaker is an electromechanical switch that opens itself when the current flowing through it exceeds a maximum, preset level. When a circuit breaker opens, or "trips," current flow is interrupted and cannot resume until the device is reset. Why does Sally include circuit breakers on the high-power side of the circuit, between the 96-V power supply and the thyristor board?

2. Suppose that 100 A of steady current flows from the 96-V battery to the power controller. If the controller circuit is 92% efficient and the motor 95% efficient, how much mechanical power is transferred to the motor wheels (neglecting bearing friction)?

3. Ohm's law states that the voltage across a resistor is equal to the current flowing through it times the resistor value. Calculate the current flowing through each of the following resistors if each has a measured voltage of 24 V across it: 1 Ω, 330 Ω, 1 kΩ, 560 kΩ, 1.2 MΩ.

4. Kirchhoff's current law states that the algebraic sum of currents flowing into a common connection, or *node*, must sum to zero. Suppose that currents of 1.2 A, -5.4 A, and 3.0 A flow on wires that enter a four-wire node. What current must flow *out* of the fourth node?

5. Kirchhoff's voltage law states that the sum of voltages around a closed path must sum to zero. Three resistors are connected in series to the 96-V battery bank. The measured voltages across two of them are equal to 50 V and 25 V, respectively.

   a. What is the voltage across the third resistor?
   b. If the first two resistors have values 100 Ω and 50 Ω, respectively, what is the current flowing through the three resistors?

6. High-power devices such as thyristors and power transistors are often mounted on metal heat sinks. A *heat sink* enhances the overall thermal contact between the device case and the surrounding air, leading to a cooler device and larger power dissipation capabilities. Heat removal is important, because excess heat can cause a catastrophic rise in device temperature and permanent failure. Examples of two typical heat sinks are shown in Fig. 7.11.

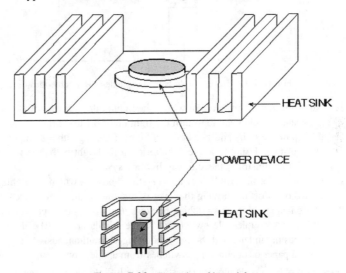

Figure 7.11 Examples of heat sinks

   Every heat sink has a heat-transfer coefficient, or *thermal resistance* Θ (capital Greek theta), that describes the flow of heat from the hotter sink to the cooler ambient air which is assumed to remain at constant temperature. This thermal flow can be described by the equation $P_{\text{therm}} = (T_{\text{sink}} - T_{\text{air}})/\Theta$, where P_{therm} is the thermal power flow out of the device, T_{sink} is the temperature of the heat sink, and T_{air} the temperature of the air.
   a. A power device is mounted on a heat sink for which Θ = 4.5 °C/W. A total of 10 W is dissipated in the device. What is the device temperature if the ambient air temperature is 25°C?
   b. A device rated at 200 °C maximum operating temperature is mounted on a heat sink. If the ambient air is 25°C and 25 W of power must be dissipated in the device, what is the largest thermal coefficient Θ that the heat-sink can have?

7. (Design problem) A switch is a mechanical device that allows the user to convert its two electrical terminals from an open circuit (no connection) to a short circuit (perfect connection) by moving a lever or sliding arm. A *switch pole* refers to a set of contacts that can be closed or opened by the mechanical action of the switch. A *single-pole, double-throw* (SPDT) switch has three terminals: a center terminal that functions as the common lead, and two outer terminals that are alternately connected to the center terminal as the position of the switch lever is changed. When one of the outer terminals is connected to the center terminal, the remaining outer terminal remains disconnected from the center terminal.
   a. Install one set of contacts in series with a 3-V flashlight bulb and two D-cell batteries. Your circuit should look something like Fig. 7.12. Verify that you can turn the light on and off by moving the switch lever.

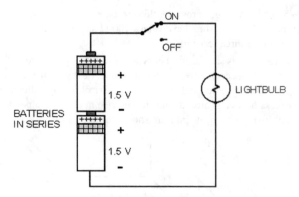

Figure 7.12 Battery, light bulb, and single-pole, double-throw (SPDT) switch

b. Now consider the problem of wiring the light in the stairway of a two-story house. Ideally, the occupants should be able to turn the light on or off using one of two switches—one located at the top of the stairs and one at the bottom. Toggling either switch lever should make the light change state. Using two D-cell batteries, a flashlight bulb, and two SPDT switches, design a circuit that illustrates the stairway lighting system.

c. Now consider the problem of a *three*-story house in which the lights in the stairwell are to be turned on or off by moving the lever of any one of three switches—one located on each floor. Design an appropriate switching network using two single-pole switches and one double-pole switch. (A double-pole switch has six terminals and can be thought of as two single-pole switches in tandem, with both levers engaged simultaneously.) You may either design the circuit on paper or obtain some switches from an electronics supply store and actually wire it.

8. (Design problem) A dc motor consists of a multi-pole electromagnet coil, called the *armature*, or sometimes the *rotor*, that spins inside a constant magnetic field called the *stator* field. In the small dc motors typically found in model electric cars and toys, permanent magnets are used to create the constant magnetic field. In larger, industrial-type motors, such as an automobile starter or windshield-wiper motors, the stator field is produced by a second coil winding.

Current is sent through the rotating armature coil by way of a set of contact pads and stationary brushes called the *commutator*. Each set of commutator pads on the rotor connects to a different portion of the armature coil winding. As the rotor rotates, contact is made by the brushes with different pairs of commutator pads, so that the portion of the armature coil receiving current from the brushes is constantly changed. In this way, the magnetic field produced by the rotating armature coil remains stationary and is always at right angles to the stationary stator field. The north and south poles of these fields constantly seek each other, and because they are always kept at right angles by the action of the commutator, the armature experiences a perpetual torque (rotational force). The strength of the force is proportional to the value of armature current, hence the speed of the motor under constant mechanical load is also proportional to armature current.

a. Obtain a small dc motor from a hobby or electronic parts store. Connect two D-cell batteries in series to the motor without regard to polarity. Observe the direction of rotation, then reverse the polarity of the battery connections and observe the results.

b. As an engineer, you are likely to encounter situations in which the rotational direction of a dc motor must be changed by a switch control. Using one of the switches from the previous problem, design a circuit that can reverse the direction of the motor using a single switch.

9. Design a switch circuit that can channel the output of a stereo receiver to one of three sets of loudspeakers.

10. Design a light circuit that incorporates three double-throw switches, batteries, and the type of small light bulbs found inside flashlights. Your circuit should indicate the majority setting of the three switches. Specifically, when two or more switches are up, one of the bulbs should light up. When two or more switches are down, the other light bulb should light up.

11. Suppose that Sally wishes to design a motor controller circuit without the benefit of thyristors. Assume the battery bank to be made up of eight 12-V automobile batteries connected in series, for a total of 96 V. Design a switching circuit that will permit voltages of various values to be connected to the motor. How many different voltage values can you achieve?

12. A variable resistor is a device consisting of two electrical terminals and a shaft that can be rotated. The resistance between the two terminals is a function of the rotated angle of the shaft; at zero angle the resistance is zero, and at 270° the resistance is at its maximum value. Using a variable resistor, design a system for measuring the level of liquid in the fuel tank of a conventional automobile. The output of your system should feed a panel display on the car's dashboard.

13. When two resistors are connected in series, the net resistance becomes equal to the sum of the two separate resistance values. When two resistors are connected in parallel, the net resistance is given by the equation $R_{eq} = R_1 R_2 / (R_1 + R_2)$. Suppose that you wish to connect together four 8-Ω loudspeakers to the same amplifier output terminals.

   a. What are the largest and smallest load resistance values that you can achieve?

   b. Devise a connection method such that the net load resistance seen by the amplifier is still 8 Ω.

CHAPTER

8

The Role of Failure in Engineering Design

Thursday afternoon . . .

The mood was somber as the engineering team filed into Harry's large office. Six months had passed since the start of the project. A raw shell of the prototype, consisting of chassis frame, outer body, drive train, electronics, and suspension system, but no finish work or paint detail, had gone out for extended test-track field trials. The tests had failed miserably. One of the front struts had broken, the power controller burned out two thyristors, and the motor had lost power suddenly in mid-travel without warning. Although the project was still theoretically on schedule, the situation looked bleak to Sally, Dave, Anne, and Keith. *My career is over*, thought Dave to himself. *It's back to square one on the strut design.* Sally was equally chagrined about the failure of her power controller. She had a fleeting thought about resigning, then realized she was being too paranoid. *Everyone makes mistakes*. Keith wondered why he was asked to come to the meeting at all, but figured it had something to do with overall quality management. Anne was the only one with enthusiasm about the meeting. She viewed the probable presence of a software bug to be a challenge and a puzzle—not unlike a mystery to be unraveled by a skilled detective. All were apprehensive about these test results, however, and expected to be "decked out" royally by Harry at today's meeting. Harry watched them file into the room and silently take their seats. Everyone assumed that he would begin the meeting by discussing the results of the prototype trials and trying to figure out what went wrong. Instead he wrote the following list on his small white marker board:

Tacoma Narrows Bridge
Hartford Civic Center
U.S.S. Vincennes

90

Space Shuttle Challenger
Three Mile Island Power Plant
Kansas City Hyatt Regency
De Haviland Comet
Hubble Telescope

Harry pointed to his list and asked, "What do these great engineering accomplishments all have in common?"

Dave was the first to recognize the common theme. "Well, we talked about the Tacoma Narrows Bridge in my mechanical design class. It fell down, or something, right?"

"Yes, it did," said Harry.

"Are these projects all examples of failure?" offered Dave.

"Yes, they are," said Harry. "Does anybody know something about the other entries on the list?" said Harry.

"Well . . . the Space Shuttle Challenger blew up on the pad," suggested Dave.

"No, it blew up during *launch*," corrected Anne.

"And didn't the Three Mile Island power plant, a nuclear one, almost blow up?" asked Sally.

"And the Hubble—its mirrors were messed up in the beginning and had to be fixed during a space shuttle flight," added Keith.

"All correct," said Harry. "The Hartford Civic Center collapsed under a heavy snow load, the U.S.S. Vincennes shot down a civilian airliner by mistake, the walkway of the Kansas City Hyatt fell down, killing scores of people, and the De Haviland Comet, a British design built in the 1950s, was the first commercial jet airplane to carry passengers. Anyone know what happened to it?" Blank stares all around. "Its fuselage developed fatigue cracks in mid-flight, leading to several crashes in the early days of civilian jet travel.

"So what's the theme here? These are all great engineering *disasters* of the past. Each involved the loss of human life or major destruction of property. And all were caused by engineering design failures. Mistakes made by engineers like yourselves who did the best they could, but had major lapses in engineering judgment or lack of prior experience directly applicable to the design task. Each of these cases has become part of engineering lore. After each occurred, similar disasters and failures were averted because engineers were able to study the *causes* of the problems and establish new or revised engineering standards and guidelines. My point is, failure is an important part of the engineering design process. Studying the compendium of classic failures helps us to learn what will work and what will not, just as studying the recent failures of our electric car prototype will help us identify hidden flaws in our design— ones that we might not have foreseen during the initial development process. Finding the cause of a failure after the fact becomes a simple game or a puzzle to be solved by an inquisitive 'Monday morning quarterback.' It's relatively easy to discover design flaws after the fact. But a truly good engineer—the kind I hope you all aspire to become—has enough on-the-job experience and understanding of classic failure incidents to spot flaws and design them out of the system *before* they lead to failure.

"The failure examples I've written on the board all had dire consequences. Each occurred once the product was in use, long after the design phase. Let's be thankful that our problems are showing up *now*, before our car has gone to market. If these sorts of failures were to show up in cars already sold to the public, the consequences would be far more serious."

It was the next part of the meeting that Harry eagerly awaited. Inside him lurked a frustrated teacher who always wondered if he would have been happier with the rigorous demands of an academic career. The junior engineers before him were not the first at Xebec to face the results of failed prototype trials. He was betting that the diversion he was about to assign them would prove instructive and also infuse some encouragement and positive energy into his distraught engineering team. He was being a good manager.

"Before we even discuss the results of our own prototype failures, we're going to attempt an exercise. I would like us to learn more about my list of classic engineering failures of the twentieth century. I'm going to assign to each person two of the entries on my list, and first thing this afternoon, you'll each present a short synopsis of your case studies. For each disaster, focus on the deficiency in the design *process*, rather than the flaw itself. You have the rest of the morning off. Go research your topics in the company library. And don't even *think* about our own prototype tests until we convene here later this afternoon. Clear?" Everyone understood.

Dave looked at his watch. It said 10:50. "Morning off?" he asked. "Sounds like we have about an hour."

"Right. Umm...well, then better get busy," said Harry. It was as close to awkward as Harry ever got. He was looking forward to the discussion that would take place later that day. Examining case studies was his hobby, and he enjoyed discussing them with new junior engineers. Perhaps he should have gone into teaching after all, he thought. Dave complied with Harry's advice and followed the team out the door to the library. *This could be fun*, he thought. *Almost like school.*
After lunch...

Harry reconvened the design team to discuss the results of their investigations. Dave, of course, was the first to go.

"OK, I was assigned to look up the Tacoma Narrows Bridge and the Hartford Civic Center. It turns out that both disasters were, in a way, related. The Tacoma Narrows Bridge, built across Puget Sound in Tacoma, Washington in 1940, was the longest suspension bridge of its day. The design engineers copied smaller existing suspension bridges and simply built a longer one. Support trusses deep in the structure of the bridge's framework were omitted to make it more graceful and visually appealing. No calculations were done to prove the structural integrity of the bridge. Because the 'tried and true' design methods used on shorter spans had been well tested, the engineers assumed that these design methods would work on longer spans. On November 7, 1940, during a particularly windy day, the bridge started to undulate and twist, entering into a magnificent torsional resonance. (Actually, some people nowadays think the bridge may have entered into a chaotic motion state.) In any event, after several hours, the bridge crumbled as if it were made from dry clay; not a piece remained between the two main center spans.

"The Hartford Civic Center also involved a type of structure that had never been built before. Its roof was made from a 'space frame' structure, like tinker toys. Multiple rods and sockets were connected together in a visually appealing geodesic pattern. Instead of performing detailed hand calculations, the design engineers relied on the 'latest' computer models to compute the loading on each individual member of the roof structure. On January 18, 1978, just a few hours after the center had been filled to capacity with thousands of people watching a basketball game, the roof collapsed under heavy snow load, demolishing the building. Miraculously, no one was hurt in the collapse."

"So what do these two incidents have in common?" asked Harry.

"For one thing," reflected Sally, "the engineers who designed the bridge and the civic center did not rely on basic intuition, gleaned from years of engineering experience, to make their decisions. In the case of the Tacoma Narrows bridge, they relied on calculations made by others, even though these calculations did not apply to the long span bridge being built. Had the engineers who designed the bridge heeded some basic scientific intuition, they would have realized that three-dimensional structures cannot be directly scaled upward without limits. In the case of the Hartford Civic Center, the design engineers relied on computer models written by programmers. One can imagine programmers who wrote the programs based on textbook formulas but who never actually built roof trusses on their own. So in one case, we have engineers who plug into formulas that don't apply, and in the other case, we have engineers who believe formulas based on computer models never fully tested on actual construction."

"I'd agree with that," said Harry. "Both basic errors in engineering *judgment*—failure to understand that knowledge of fundamentals and informed intuition must always precede computer modeling and formula plugging." He paused for a moment to let the group digest his words. "Would anyone like to share another disaster?"

"The Space Shuttle Challenger," offered Keith. "My dad clearly remembers the day it blew up. It happened during launch on a cold day in January at Cape Kennedy (Canaveral) in Florida. After weeks of investigation, NASA determined that the problem was a set of o-rings used to seal sections of the multisegmented booster rocket. Those seals were never designed to be operated in cold weather, and on that particular day—it was about 28°F, very cold for Florida—the frozen o-rings were too stiff to properly seal the sections of the booster rocket. Flames spewed from an open seal during power up and ignited a fuel tank. The entire spacecraft blew up, killing all the astronauts on board.

"In this case, we have engineers relying on a 'standard' design technique: using o-rings to seal adjacent cylindrical surfaces. This time they used them on a structure much larger than any that had been tried before. The cold temperature brought the seal to its limit, and it failed. It was the worst space disaster in U.S. history."

"But there was another dimension to the failure—something that did not influence the previous two failures that Dave discussed. Keith, did you find out *why* the booster had been built in multiple sections, requiring o-rings in the first place?"

"Sure did, boss." (Harry hated being called "boss.") "It was a *political* decision. Common sense would have had the boosters built in one piece without any o-rings at all. But a senator from Utah lobbied heavily to have the contract for constructing the booster rockets awarded to a company in his state. It was not physically possible to transport a large, one-piece booster rocket all the way from Utah to Florida over existing rail lines. Trucks were too small, and there were certainly no ships that could do the job. Utah is, after all, a land-locked state in the middle of the United States. The decision by NASA to award the contract to a Utah company resulted in a multisection, o-ring sealed booster rocket whose smaller pieces were easily shipped. A sad piece of ethics, I'd say."

"I'm not sure I'd call it a breach of ethics," said Anne. "It wasn't as if the Utah senator or engineers maliciously designed a substandard product. After all, the sectioned booster had worked flawlessly on many previous shuttle flights that were not launched in below-freezing temperatures. It's just that, by putting more weight on a political element of the project, rather than pure engineering concerns, the engineers were forced into a less-than-desirable multipiece design that had never been attempted before."

"The lesson to be gleaned from this last example," added Harry, "is that first-time designs often betray previously unknown flaws after an initial period of successful use. Design flaws eventually show up because the operating environment changes."

"Couldn't we attribute such a failure to plain old statistics?" asked Dave. "I mean, if something is bound to fail, it *will* fail sooner or later. After it's failed, the engineers in charge of designing the product can fix what's wrong in future releases. That's why I'll never buy a new car in the first model year. I like the model to have been out a few years to let other people discover the car's bugs and weaknesses."

"Nevertheless, new products, our electric car included, will always keep coming out," said Harry. "That's what technological progress is all about. It's our job as design engineers to identify as many of the bugs and weaknesses as possible through testing, retesting, and more testing under all sorts of operating conditions. So in a way, we should look upon our recent unsuccessful prototype tests as an excellent opportunity to discover and weed out the bugs before the car goes commercial."

There was a brief lull in the conversation. Harry spoke again. "Keith, what was your second assigned case study?"

"I can see why you assigned me these particular cases," said Keith. "My second was the Kansas City Hyatt accident. If any of you have ever been into a Hyatt hotel, you know that their internal architectures are rather unique. The typical Hyatt hotel has cantilevered floors that form an inner trapezoidal atrium, and the walkways and halls are open, inviting structures. There's nothing quite like the inside of a Hyatt. In the case of the Kansas City Hyatt, the original design called for a two-layer, open-air walkway to span across the entire lobby in midair, from one balcony to another. During one particular event—a party of some sort that took place not long after the hotel opened in 1981—the walkway had a lot of people on it, and they were all dancing in time to the music. The weight and rhythm of the load of people, perhaps in resonance with the walkway, caused it to collapse suddenly. Over one hundred people died. It was a horrible disaster that will be remembered forever in the history of hotel management. Although the hotel eventually reopened, to this day the walkway has never been rebuilt.

"In order to explain how the walkway collapsed, I need to draw a picture. Here's a sketch of the skeletal frame of the walkway as specified by the design engineer." Keith drew the sketch of Fig. 8.1 on the marker board.

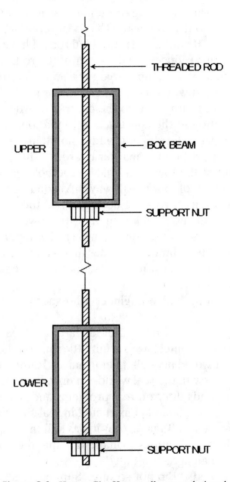

Figure 8.1 Kansas City Hyatt walkway as designed

"Each box beam is held up by a separate nut threaded onto a single suspended steel rod. The rated load for each nut-to-beam joint is intended to be above the maximum weight encountered during the time of the accident. Anybody see anything wrong with this picture?"

"Looks OK to me," said Dave. "The rating information must have been faulty."

"I agree," said Sally. "The data used by the engineer must have been wrong."

"Well, that's not the main problem here. It really has to do with the construction method, not the structure itself," said Keith.

"Wait," interrupted Anne. "I see what the problem was. There's no way that structure could ever have been built. If you look at the steel rods, they're threaded all the way from the lower end to the middle, where the upper walkway is supported. It looks like a distance of twenty feet or so. That would mean about forty feet of rod to be threaded per double hanging rod, and there must have been about ten or fifteen rod locations holding up the walkway. The rods were probably made from plain steel rod having no threads. If you've ever threaded rod—I have, on a summer co-op once—you know that threading reduces the diameter of the rod. There's no way that you can get a nut to the middle of a rod unless you thread the entire length. Even with an electric threading machine, it would have taken *days* or *weeks* to thread all those rods."

"Right," said Keith, "and so the contractor who actually built the walkway proposed a modification to the construction so that only the very ends of the rods would have to be threaded. Here's the modification that was proposed." Keith drew the revised sketch of Fig. 8.2 on the marker board. "What was wrong with the modification was that the nut at the lower end of the rod holding the upper walkway now had to support the weight of *both* walkways. A good analogy would be two mountain climbers hanging onto a rope. If both grabbed the rope simultaneously but independently, the rope could hold their weight. If the lower climber grabbed the ankles of the upper climber instead of the rope, however, the upper climber's hands would have to hold the weight of *two* climbers. Under the full, or maybe excessive, load conditions of that day, the weight was just too much, and the joint gave way. Once one joint failed, the rest quickly followed."

"That was some naive engineer who specified single rods with twenty feet of threading," said Dave.

"Well, maybe, maybe not," offered Harry. "It's easy to imagine how a design engineer, or perhaps an architect, who spent little time on the construction floor, could make such a mistake. I would not call it being naive. Rather, I'd call it being inexperienced—lacking experience with the real world. In my day, we called it 'seasoning'—the process of getting your hands dirty on real problems and real construction—getting a feeling for how things are made and constructed in real situations.

"That may be," said Dave, "but it was still the responsibility of the engineer who signed off on the modifications to make sure that the revised design provided adequate strength. Can we all agree on that?"

Sally disagreed. "I'm not so sure. Suppose the person signing off on the design change was a junior engineer under the supervision of a senior mentor. The junior engineer might not understand the ramifications of the design change and assume it to be routinely trivial. It would have been the responsibility of the senior engineer to adequately communicate the critical nature of the structure as designed."

"And what about leaving plenty of room for safety margins?" added Anne. "From what I know of structural design, it's common practice to leave *at least* a factor of two safety margin between calculated maximum load and the structure as actually built. The safety margin allows for inaccuracies in load calculations due to approximation, random variations in material strengths, and small errors in fabrication. Had the walkway included a safety margin of a factor of two or more, the double stressed

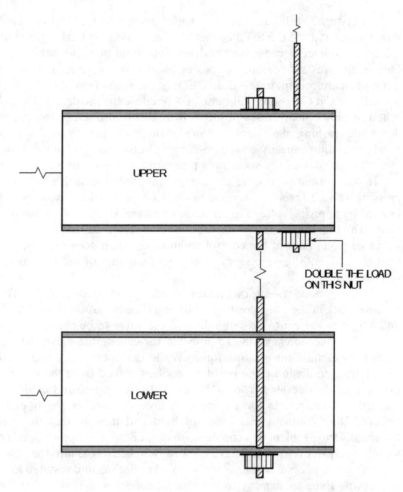

UPPER

DOUBLE THE LOAD
ON THIS NUT

LOWER

Figure 8.2 Kansas City Hyatt walkway as actually built

joint on the walkway might not have collapsed, even given the as-constructed modification. I'd put the blame for that error on the head of the original architect."

The arguments went back and forth across the room, reaching no clear conclusion about who was to blame for the accident. There was unanimous consensus, however, about how such errors might be prevented in the future: A design effort should include workers from all phases of construction in the design process. Steps must be taken to ensure adequate communication between all levels of employees, and design choices should include far more than minimal safety margins where public safety is at risk.

When they had exhausted all opinions on the Hyatt and Space Shuttle disasters, Harry said, "OK, who's next?"

"My turn?" offered Anne. "I was supposed to look up the Three Mile Island Power Plant and the U.S.S. Vincennes incidents. As best I can figure out, these two accidents are related because they both involved poor human-machine interfaces. Three Mile Island was a large nuclear power plant in Pennsylvania. It was the site of the worst nuclear accident in the United States; the second worst in the world after the total meltdown at Chernobyl, Ukraine. Fortunately, the incident at Three Mile Island resulted in only a near miss at a meltdown. But it also led to the permanent shutting down and trashing of a billion dollar electric power plant and significant loss of electrical generation capacity on the power grid in the eastern United States.

"On the day of the accident, a pressure buildup occurred inside the reactor vessel. It was normal procedure to open a relief valve in such situations to reduce the pressure to safe levels. The valve in question was held closed by a spring and was opened by applying voltage to an actuator coil mechanism—an electromagnet, essentially. The designer of the electrical control system had made one fatal mistake. The indicator panel back in the control room lit up when power was applied to the valve actuator coil, but it gave no indication about whether or not the valve itself had *actually* opened.

"You guessed it—after a pressure relief operation, the valve at Three Mile Island became stuck in the open position. Although the actuation voltage had been removed and a light in the control room indicated the valve to be closed, it was actually stuck open. The mechanical spring responsible for closing the valve did not have enough force to overcome the sticking force. While the operators, believing the valve to be closed, tried to diagnose the problem, coolant leaked from the vessel for almost two hours. Had the operators known that the valve was open, they could have taken other corrective measures. In the panic that followed, however, the operators continually believed their control panel indicator light and thought that the valve was closed. Eventually the problem was contained, but not before a rupture nearly occurred in the vessel. Such an event would have spewed radioactive gas into the atmosphere, exposing millions of people to dangerous doses of radiation, and resulted in a complete core meltdown. Even so, damage to the reactor core was so severe that the plant—which had cost billions to build—was permanently shut down. It has never reopened.

"My other incident involved the U.S.S. Vincennes which was also plagued by a poor machine/human interface. The Vincennes was a U.S. missile cruiser stationed in the Persian Gulf during the Iran/Iraq war. On July 3, 1988, while patrolling the Gulf, the Vincennes received two IFF (Identification, Friend or Foe) signals on its Aegis air defense system. Aegis was the Navy's complex, billion dollar, state-of-the-art information processing system that displayed more information than any one operator could hope to digest. Information saturation was commonplace among operators of the Aegis system. The Vincennes had received two IFF signals, one for a civilian plane and the other for a military plane. Under the pressure of anticipating a possibly impending attack, the overstimulated operator misread the cluttered radar display and concluded that a warplane was approaching the Vincennes. Repeated attempts to reach the non-existent warplane by radio failed. In the charged atmosphere of the events taking place during that time in history, the captain concluded that his ship was under attack and made the split-second decision to have the airplane shot down. Two hundred and ninety civilians died needlessly. My question: Was this an operator error or an engineering error?"

Sally was the first to react. "I see the connection between these two cases. Both involved poor human/machine interfaces. In the first case—Three Mile Island—the operators correctly assumed that the information they were receiving was accurate, while in reality it was not because the designing engineers had not thought about the environment in which the system would be used. The critical test of such a system, of course, would be during an emergency when the need for absolutely accurate information would be acute. The power plant's control panel provided the key information by inference, rather than by direct confirmation.

"The same could be said of the Vincennes incident. I remember reading about it in the papers. The Navy officially attributed the accident to 'operator error' by an enlisted crew, but in some circles the blame was placed on the engineers who designed the system. Under the stress of possible attack and deluged with information, the operator simply could not cope with the ill-conceived machine/human interface designed by the engineers."

No one had any additional comments to add to Sally's synopsis. Each had understood why Harry had put these examples on his list. The interface between the driver and the Xebec electric car had many similarities to both the Three Mile Island control panel and the Aegis system.

Harry broke the silence. "The lesson to be learned here: critical information will be needed most during crisis situations. It had better be accurate, up-to-date, and easy to interpret. The weakest link in any electromechanical system is bound to be the human interface." He was sure the team had understood the message.

Last on the list of participants was Sally. "I was assigned to look up the De Haviland Comet and the Hubble Telescope. As you probably know, of the many problems that plagued the Hubble, the most famous was its improperly ground mirrors. They were distorted and had to be corrected by the installation of an adaptive optic mirror that compensated for the aberrations. The repairs were carried out by a NASA Space Shuttle crew. But this error was attributed to sloppy mirror fabrication. The real engineering error that fits into our common theme is the problem with the Hubble's solar panels. When they were deployed in the environment of space, they were alternately heated and cooled as they moved in and out of the earth's shadow. The resulting expansion and contraction cycles caused the solar panels to flap like the wings of a bird. Attempts to compensate for the unexpected motion by the spacecraft's computer controlled stabilizing program led to a positive feedback effect which only made the problem worse.

"My second case is probably one you've not heard of before today. The De Haviland Comet was the first commercial passenger jet aircraft. A British design, the Comet enjoyed many months of trouble-free flying until several went down in unexplained crashes. Investigations of the wreckage suggested that the fuselages of these planes had ripped apart in mid-flight. It turns out that no one had foreseen the effects of the numerous pressurization and depressurization cycles that were an inevitable consequence of takeoffs and landings. Before jet aircraft, airplanes were not routinely operated under pressure. In the case of the Comet, the locations of the rivets holding in the windows developed fatigue cracks which, after many pressurization and depressurization cycles, grew into large, full-blown cracks in the fuselage."

"And the common theme here is obvious," Harry interjected. "In both these cases, the design engineers had not thought about the environment under which the finished product would be used. Content with laboratory tests that did not mimic these actual operating conditions, the engineers were lulled into a sense of security about the soundness of their designs. The moral: always test a design under the most realistic conditions possible. Never assume that environmental conditions will not affect performance or reliability. In the case of our car's prototype tests, that's exactly what happened. The prolonged exposure to real road conditions caused unforeseen failures. It will be our job to correct these design flaws and test the vehicle again and again before the car makes it to the marketplace."

Harry and the team discussed the case studies for about another hour. It was clear to Harry that his plan had worked. The four engineers appeared renewed and buoyed with confidence that engineering failures are part of the design process. Failures on the test track were normal. They viewed the resolution of these problems as their next challenge. In the weeks that followed, they returned to their work on the electric car project with renewed vigor.

REFERENCES

H. Petroski, *To Engineer Is Human: The Role of Failure in Successful Design*. New York: Vintage Books, 1992.

E. S. Ferguson, "How Engineers Lose Touch," *Invention and Technology*, vol 8 (3), Winter 1993, pp. 16-24.

EXERCISES

Look up and write a synopsis of each of the following classic engineering failure incidents:

1. General Electric rotary compressor refrigerators (1990)
2. Exxon Bayway refinery, Linden, New Jersey (1990)
3. Green Bank radio telescope (1989)
4. Quebec City bridge (1907)
5. American Airlines DC-10 (1979)
6. Skylab (1979)
7. Interstate 95 Bridge, Mianus River, Connecticut (1983)
8. Point Pleasant Bridge, Ohio River, Ohio-West Virginia (1967)
9. Big Ben, London (1976)
10. Liberty Bell, Philadelphia (1835)
11. Alexander L. Kielland oil platform, North Sea (1980)
12. Korean Airlines Flight 007

9

Learning to Write as an Engineer

Friday morning, 9:00 AM . . .

Fran, a technical writer from Publications, was talking at the front of the conference room. Harry had arranged for the entire project team to attend a day-long writing workshop. He hadn't been very pleased with the memos and internal documents that the electric car team had been producing and had issued an edict that everyone improve their writing skills . . . or else!

Or else what? wondered Dave as he focused on Fran's presentation. He was not complaining, though. He was not a good writer, he knew it, and he was not especially proud of it. Technical writing was simply one skill that had never occupied central stage in any course he had taken in college. The English prose style stressed in his freshman writing course had not been very useful for writing the engineering documents and memos now required of his job. Several employees from other divisions were also present. Fran's writing course was a biannual event upon which the company had come to rely.

Fran put up an introductory slide with her name and credentials. "We're going to cover a lot of ground today, and you'll do more than just listen to me talk. In this workshop, you'll get a chance to practice writing using the techniques that I am going to present. It's simply not possible to learn how to write solely by reading instructions. Writing, like sports or art, requires practice, practice, and more practice.

"If everyone's here, we'll get started. Before we begin, I thought I'd give you an overview of today's workshop. Here's a summary of the kinds of writing documents that we'll look at this morning." Fran put up the following slide:

Xebec Research and Development
Writing Workshop
Simple Memos
Short Reports
Technical Instruction Manuals
Proposals and Journal Papers

After going through the list and making a short comment on each, Fran continued. "The first topic on the agenda is the writing of simple memos. As our first example, let's suppose that you want to write a simple memo to your boss describing the results of some engineering tests. Here's a summary of what such a memo should contain." Fran put up the following slide:

Writing a Simple Memo
Determine ahead of time:
- The recipient of the memo
- The purpose of the memo
- The content of the memo

"Writing a simple memo requires that you identify three things: the recipient, the purpose, and the content. Knowledge of the *recipient* is important because it sets the tone of the memo. Will the memo be formal, informal, alarming, humorous? Deciding upon the *purpose* of the memo helps you formulate its tone and state its objectives. Will your memo provide information, request a response, ask for input, or give instructions? Determining the *content* of the memo will dictate how its body will be organized. For a memo in a technical setting, such as the internal memos you're likely to write for this company, the ideas and information must be organized in clear, concise, and logical form.

"When you engage in a face-to-face conversation, you can modify your communication approach on the spot, depending on the person's reaction. But a written memo will be read in private, and you will not have the benefit of witnessing your recipient's reaction. A written memo must elicit the desired reaction by being carefully crafted.

"One of the most common mistakes in memo writing is to assume that messages sent electronically via e-mail do not have to be prepared with the same care as other written documents. Nothing could be further from the truth. The recipient of an e-mail memo will read it in private, just like a written memo, and may treat it with the same formality as a memo received on paper. More importantly, an e-mail memo can be copied to large numbers of people instantly, greatly multiplying its exposure. The same techniques applied to paper memos should also be applied to e-mail messages. Many people also print out e-mail messages, effectively converting them to paper form, so the distinction between paper and e-mailed memos is often blurred anyway." Fran circulated a handout that contained the information on her next slide:

Guidelines for Writing a Good Memo
- Begin with a header containing: To, From, Subject, and Date
- Identify the purpose of the memo in the first (or second) sentence
- Clearly present the body of information
- Sign the memo

Fran's handout included a second sheet with a sample memo on it. "I'd like you all to look at this memo which was written by Harry Vigil several years ago before he became manager. With his permission, I've saved it to use in my workshops. Note that Leo was Harry's manager at the time of writing. When you've finished reading the memo, we'll discuss it." The workshop participants studied the memo which read as follows:

To: Leo Arbanian
From: Harry Vigil
Subject: Upcoming ASME Conference

January 12, 1990

Leo,

    As you may know, the American Society of Mechanical Engineers is holding a conference on lightweight composites at the end of June in Dayton, Ohio. I think that Xebec should send someone to attend this meeting. Over the past several years, composites have shown promise as a viable alternative to steel or aluminum. They combine the strength of the former with the light weight of the latter. It's time we learned more about these important materials.

    The conference will be held at the Dayton Buena Vista. I've spoken with our travel agent and found that flights will cost between $200 to $300 round trip. Hotel will be $62 per day, and the conference registration fee is $180. Please let me know what you think.

Harry

"So," asked Fran to the group, "tell me what you think of the memo."

"Seems well written," offered Keith. "Harry began the memo by explaining the upcoming conference and concisely presented his data about how much it would cost him to make the trip."

"I agree," said Dave. "Seems like a good memo to me."

"Well," explained Fran, "Harry never got to go to the conference, even though Leo was convinced of its value by his memo. Does anyone see why?" The group pondered Fran's question. Sally raised her hand.

"Because he never clearly stated the purpose of his memo?"

"Exactly right," said Fran. "Harry presented his data, did his research, but never stated explicitly that it was *he* who wanted to attend the conference. The memo brought the matter to Leo's attention, and it was Leo himself who wound up going to the conference! This memo would have been *much* more effective had Harry begun it with the simple sentence, 'I would like permission to attend an upcoming ASME conference in June.' The lesson learned is, always state the ultimate purpose of your memo in the first sentence. So doing adds immense clarity and avoids misunderstandings about the memo's content."

When she was finished, Fran passed out another sample memo. "I'd like you next to read the memo I've just handed out. I saved it from a workshop that I gave several years ago. The memo, written by someone named Olga Galli, deals with the rewriting of a mainframe Unix software program called the Universal Information System. The fictitious task involved rewriting the program to run under Microsoft Windows™. The software program was used by the client to keep track of customer charge card records. The assignment in my workshop was to write a memo to the boss summarizing the results of a recent exploratory visit to the client. Prior to the writing exercise, the details of the fictitious visit and project were distributed.

"One quality of good writing," continued Fran, "is its ability to correctly convey the desired information, including all of its subtleties, to the reader. What I'd like to do is go around the room and have each person comment on what's good and bad about this memo. By going through this exercise, we'll be able to see how well this memo accomplishes its task. We'll also be able to quickly zero in on its deficiencies. We should then be able to arrive at a consensus of what goes into a well-written document." The memo handed out by Fran read as follows:

Universal Information System
Sample #1 (original version):

Customer Meeting

On the 17th of September, my group met with the customer. We were allowed to use a user's ID in order to bring up his account record on the UIS. The customer crossed out the user's name on the sheet containing his information but it turned out that the UIS listed his account with both his name and social security number. My group made a record of how the system was set up in order for us to have the same heading in our program.

I learned some of the commands that are used on the UIS Galaxy system. The command TR14 finds the user's in question account and displays it on the monitor. The TR33 command displays all transactions which the user made since the specified date. Last, but not least, the TR35 command displays all payments made since the given date. It turned out that this command was a recent one designed and implemented by one of the fellow employees who recently left the company. In our program we will assume that this command will be used by the customer, even though there is doubt that no one would keep it up to date except for the fellow employee whose stay there is not guaranteed.

Meeting with the customers at their office made it clearer what they expect of the product. They want the product easy to use. They want the account summary sheet to highlight all the transactions not yet paid for. In all, I learned several commands and got some exposure to the UIS system. This was a profitable meeting.

"Let's first make a list of the key points of information included in the memo." Fran wrote on an overhead projector as various participants suggested the following items:

- Opportunity to see demonstration of UIS system
- Customer erroneously allowed us to see the identity of a sample account holder
- System is called "Galaxy"
- UIS system includes several commands:
  - -TR14: Customer account
  - -TR33: Transactions display
  - -TR35: Payment history
- TR35 command designed by previous Xebec employee
- Meeting was useful
- Customer wants unpaid transactions highlighted

"OK. Comments?"

Sally was first to respond. "For one thing, the order in which the information is presented is entirely random. It's almost as if the writer just wrote ideas down in whatever order they came."

"Good," said Fran, "I agree. No logical progression of ideas. Other comments?"

Dave, who had been thinking about lunch, suddenly popped up his hand. "That stuff in the first paragraph about seeing the account holder's identity—I mean, who cares?"

"I would care if I were the customer," suggested Anne. "Those records are a matter of privacy."

"No, *that's* not what I mean," answered Dave, putting emphasis on the word "that." "I mean, the reader of the memo, Olga's boss, could probably care less about the fact that the customer made the mistake in front of a Xebec employee. It's just not relevant to the memo, and especially should not appear in the *first* paragraph which, as Fran has noted, should always include the purpose of the memo."

"OK, I'd certainly agree with that. Sorry for my misinterpretation," replied Anne. Dave was fascinated, not annoyed, by the dual meaning of his words. "I also see," added Anne, "that the writer did not include a heading and did not sign the memo."

"Good. Other comments?" Fran continued her prompting of the participants.

"Well," offered Keith, "there's no mention of who the customer is or who the person was that gave the tour and demonstration. Similarly, there's no mention of who the author is."

"And there's no mention at all of what 'UIS' stands for—a bad assumption on the part of the writer that the reader knows the meaning of the mnemonic. She also just used the term 'Galaxy'—I assume that's a subsection program within the UIS system—without explaining what that term means."

"Yes, true," offered Fran. "For your information, the customer was called 'Boulton Industries' and the contact someone named Connie Donaldson. Another problem with this memo is that it contained no preamble, introductory sentences, or auxiliary notes. Each fact is presented coldly with no accompanying explanation of how it fits into the context of the memo. It's almost as if the writer had our list of facts in her head and proceeded to simply regurgitate them on paper. Obviously, this memo underwent very little editing and re-writing.

"Your observations about this memo are all good ones," continued Fran. "Let me summarize them." She wrote the following list on the overhead:

- No logical progression of ideas
- No heading or signature
- No mention of sender or recipient
- No mention of customer's company or name
- Irrelevant information (about customer's error during demonstration)
- No mention of the purpose of the memo
- No explanation of "UIS" mnemonic or the term "Galaxy"
- No preambles, introductory sentences, or auxiliary notes

"I'd like now to show you a revised version of the memo which I personally rewrote for this workshop. I think you'll see that it corrects most of the deficiencies on our list." Fran handed out another sheet of paper:

Sample #1 (revised version):

To: Company President
From: Olga Galli
Subject: Summary of customer meeting at Boulton Industries

On September 17, the Xebec project group met with our company's customer, Ms. Connie Donaldson, in the office of Boulton Industries. The meeting provided us with an overview of the customer's existing Universal Information System (UIS) and introduced us to the procedures commonly used in Ms. Donaldson's office. We worked on the actual UIS system and dealt with real account records. This memo summarizes what we have learned.

The UIS system contains an application program called Galaxy which allows the customer to view account records. The record of a given account holder is accessed by entering his ID number in to the Galaxy system at the menu prompt. The account data can then be viewed in a variety of formats useful to the customer.

Several user programs, called "TR screens," reside within Galaxy and help the user to display account information. These screens are used routinely by Boulton employees in the processing of monthly bills and in servicing customer call-in inquiries. During our meeting, we recorded the display headings of each important TR screen so that we can create the same headings on our PC-based Windows system.

As part of our tour, Ms. Donaldson demonstrated some of the commonly used TR screens. One command, called TR14, finds a given account record and displays it as a standard mailing form on the monitor. A second command, called TR33, displays all transactions that have occurred since a date specified by the user. A third command, called TR35, displays the history of all payments since the user-specified date. This last command in UIS was written for Boulton by a previous Xebec employee. When writing our program, we should assume that this command will be used by Boulton to process records and integrate the feature into our program.

I feel that we learned enough from our meeting to enable us to proceed with the Boulton UIS project. If necessary, we can return to Ms. Donaldson's office at a later date to obtain more information.

Sincerely yours,

Olga Galli
Xebec Software Division

Everyone agreed that Fran's version of the memo was much improved. She passed out another sheet. "Now I want you to look at a second memo which I culled from one of my previous workshops. This memo was based on the information list shown in the following slide. The list was intentionally sketchy and disorganized, and the assignment was to write a memo to Harry explaining its contents." Fran put up the following slide:

-Initial loading tests completed
-Samples were composites and steel
-Same shape for each; steel machined, composites molded
-Initial difficulties fitting samples into test machine
-Made a holding jig to allow testing
-We should go with composites at slightly increased diameter.
-Data:

DIAMETER	COMPOSITE	STEEL
0.25 in	245 lbf	321 lbf
0.375 in	1644 lbf	1790 lbf
0.50 in	3021 lbf	3229 lbf

The memo that Fran handed out read as follows:

To: Harry Vigil
From: Fred Barber
Subject: Test on Composite Materials

Harry,

We've completed the initial loading tests on the samples made from composites and steel. The samples had the same shape and the steel was machined and the composites molded. We had some initial difficulties fitting the samples into the test machine but finally made a holding jig to allow testing. I think that we should go with composites at slightly increased diameter.

Here is the data:
    0.25 in diameter: composite 245 lbf, steel 321 lbf
    0.375 in diameter: composite 1644 lbf, steel 1790 lbf
    0.50 in diameter: composite 3021 lbf, steel 3229 lbf

Sincerely yours,

Fred Barber

"Comments?" asked Fran. "How do you think Fred did?"

"Well," offered Anne, "his heading looks fine and he did clearly state his objective in the first sentence. At least, I think he did."

Sally raised her hand. "Actually, I don't think he did so at all. The ultimate purpose of the memo was to make a recommendation to Harry that composites be used—I presume on our electric car project. A better opening would have been, 'We've completed the initial loading tests on the samples. Based on our test results, I'd like to recommend that we go with composites.'"

"Good, Sally," said Fran, "I would agree with that comment. Are there others?"

"The memo seems rather disorganized to me," said Keith. "I mean, it's organized in the sense that he's followed almost verbatim the ordering of ideas as presented in your slide, but those ideas were not particularly well organized to begin with. The memo reads disjointly, as if it's a poorly edited movie."

"I agree," said Anne. "His sentences are choppy and don't flow from one to another. He should have presented the numerical data in more concise, tabular form. Also, I don't think the content of his memo is especially clear. He didn't describe the purpose of the tests, didn't go into detail about how they were performed, and didn't describe the test samples other than to say that they were a mix of composites and steel. He should have at least given their dimensions, the composition of the composites, and how many of each type were tested."

Dave's hand shot up. "And he left out the date!"

"Very good, Anne, Dave," said Fran. "Good points, all. That seems like a reasonable number of comments for now. Now, here's what I want you to do. It's your turn to do the rewriting. Each of you should revise Fred's memo, taking into account all of the comments you've made so far. Put down your name as the sender so we'll know whose revision is whose. When you're finished, I'll make overheads and put up your revised versions for further discussion. Don't be afraid to rewrite several times. You should revise the memo until you are happy with it."

The group got right to work. The exercise took about twenty minutes. When they were done, Fran took their final handwritten versions to a copy machine in the next room, made overheads, and showed them one by one to the group. Dave's revised memo read as follows:

> To: Harry Vigil
> From: Dave Jared
> January 16, 1997
>
> Harry,
>
> The purpose of this memo is to recommend composites over steel as a material for the new electric car frame. We have completed the initial loading tests on samples of composites and steel, and the results are summarized below. Note that we initially had some difficulty fitting the samples into the test machine, but eventually made a test jig that solved the problem.
>
> | | FORCE | | | |
> |---|---|---|---|---|
> | Diameter: | 0.25 in | 0.375 in | 0.50 in | 0.625 in |
> | Composite: | 845 lbf | 1644 lbf | 3021 lbf | 4229 lbf |
> | Steel: | 1421 lbf | 2790 lbf | 4310 lbf | 5541 lbf |
>
> Sincerely yours,
>
> Dave Jared

"Discussion?" asked Fran.

"I got the date in," noted Dave.

"Well," offered Keith, "you did a much better job of organizing the data into a table. It's still not clear from your memo, however, what the details of the tests were or why you performed them. And your simple title for the table, 'Force,' is a bit sketchy."

"And," added Sally, "the first paragraph seems forced, as if you were following Fran's instructions to cite the purpose of the memo in the first sentence."

"Which I was," noted Dave, "but I see what you mean."

"I think some of these items are much improved in Anne's version of the memo," said Fran. "May I put yours up for discussion, Anne?" A nod from Anne. Her revised memo read as follows:

> To: Harry Vigil
> From: Anne Richards
> January 16, 1997
>
> Harry,
>
> The mechanical group working on the Plugmobile electric car project has just completed tests on samples of the composite and steel materials we are evaluating for the main structural members of the frame. Based on our test results, our group recommends that we choose com-

posites of slightly larger diameter for the structural materials. The details of the tests are described below.

Samples of machined steel and molded composites were all fabricated in the shape of standard tensile test specimen bars having a variety of diameters in the range 0.25 inch to 0.625 inch (ASME specification 246). These samples were stressed to the breaking point in our Instron Model 2000 test machine. Although we had some initial difficulties in fitting the samples of diameter other than 0.5 inches into the test machine, we eventually made a holding jig to accommodate testing of these odd sized samples. The numerical data from our test results is summarized in the following table:

BREAKING FORCE UNDER TENSION

Diameter (in):	0.25	0.375	0.50	0.625
Composite (lbf):	845	1644	3021	4229
Steel (lbf):	1421	2790	4310	5541

As you can see, our minimum targeted breaking strength of 4000 lbf can be met with either a composite rod of 5/8 inch or a steel rod of 1/2 inch. Given the much lighter weight of the composite material, however, its strength to weight ratio is much higher than that of steel.

Anne Richards

Sally was the first to comment on Anne's memo rewrite. "I like this version much better. She's mentioned the purpose of the memo in the second sentence, rather than the first, but it works well here because the memo still reads and flows nicely. The first sentence serves as a preamble to the real purpose of the memo which is first revealed in the sentence that begins, 'Based on our test results...'"

"Good," said Fran. "Any other thoughts, class?" Her use of the word "class" caused Dave to imagine for a moment that he was back in school.

"I like the way she's presented her table," he said. "Instead of putting the units next to each entry, she's written them once next to just the entries on the left-hand side. It makes for a much neater looking display of data. Also, she's elaborated on the details of the test—although obviously she's just made up a lot of the details for the purpose of this exercise—but the added information makes the memo much more informative. Anyone reading it, Harry included, would immediately know the context in which it was written."

"That's actually an important point," added Fran. "In writing the memo, Anne may think that Harry will immediately understand its context because her primary focus has been the taking of data and the testing of the materials. But Harry, as a manager, is likely to be juggling dozens of projects and details, and he will probably welcome a memo that first refreshes his memory about its context and background. In addition, he may find himself reading the memo at some future time when the background details of the memo, if omitted, might have faded a bit from his memory. Or he might decide to forward the memo to someone else who is unfamiliar with the details that Anne has correctly provided. Do we agree that Anne's memo is a much improved version?" Nods from all around. The group discussed several more memos, then took a break. The workshop included pastries, muffins, coffee, and juice.

After the break, Fran continued on to another topic. "We'll now move on to the writing of short reports. As employees of Xebec, you are likely to be engaged in this activity in the near future. Most of our contract work requires that we write periodic progress reports. The task is not unlike the lab reports you probably had to write in some of your college courses. The typical format is canonical: Introduction, Data, Analysis, Conclusion. For the purpose of this workshop, we'll define a short-report document as being between, say, two and twenty pages long—not so long that major organizational steps must be taken to write the document, but long enough so that the style is different from that of a simple memo. With your permission, the technique I'd like to take is to critique a report recently written by someone from this group. I'll pass it around, and we can all discuss it. Then we'll take an hour break, during which time you'll attempt to rewrite the sample report. As I mentioned at the beginning of the workshop, one learns writing by practicing, not by listening to me talk about it."

Great, thought Keith, *we can retire to our offices. I can check out my e-mail before we get started to make sure there are no crises with those production line tests.*

"...and you should stay in this room," continued Fran. "One of the most important lessons about writing longer documents is that you *must* devote an uninterrupted block of time to the task. It's simply not possible to write well if you are distracted by phones, e-mail, or people coming to talk to you. We'll sequester you in this room so you can get your writing task done. If it were up to me, the company would issue "Writing in Progress—Do Not Disturb" signs for all employees to post at their doorways and cubicle entrances. The company's writing productivity would increase immensely.

"Before we begin, let's go over the basic components of a good report. These features are meant to be guidelines only. Obviously, one has to alter them to fit the task.

Introduction

"The preamble, or introduction, cites the purpose of the document and the reason for its being written. Unlike the simple memo, where one attempts to cite the purpose in the first or second sentence, the introduction of a longer report can take more time to explain its purpose. The reader presumably has more time to read the document and can be eased into its contents. A person pressed for time and merely skimming the document will tend to skip the introduction anyway, so it's best to write the introduction under the assumption that it will be read in its entirety only by serious readers.

Data Section

"The data section includes the technical material relevant to the report. It's important to explain why each set of data is presented, how it was obtained, and what bearing it has on the main purpose of the report. A report is likely to stay on the shelf and its data section used later for reference purposes. Hence it's important to present data completely and thoroughly, but in a way that is not confusing and is easily digested at a glance.

"The analysis section is where you show how the data supports (or refutes) the claims you make in the report. Mathematical calculations belong in this section, as do plots and charts derived from the basic data. In some cases, particularly in reports that deal with design work, the analysis and data sections appear in reverse order—first the analysis of the device is presented, followed by data on tests that show whether the device meets the expectations or predictions of the analysis.

Conclusion

"Finally, the conclusion is used to summarize the claims, results, and observations included in the report. The conclusion section should be written to serve the needs of the browser or skimmer—someone who may not have time to read the whole report but needs to be familiar with its content. The conclusion should be a stand-alone section that summarizes all the key points of the report."

Fran led a discussion of the report, then sat everyone down for about an hour to tackle the rewriting task. Her well-articulated guidelines proved especially helpful. After a second round of discussions, the group broke for lunch.

After lunch, the team reconvened. The last topic of the day concerned the writing of long documents—reports, instruction manuals, and technical journal papers. "Since these documents are usually lengthy ones," said Fran, "it's impractical for me to show examples and have you rewrite them. Instead I'll just talk about techniques that you can use anytime you need to write such a document.

"In general, writing a report, proposal, or paper of substance takes a long time, from hours to sometimes days. It takes persistent concentration over an extended period of time to get into a creative writing mode. Your brain must sort ideas, arrange their flow, and commit them to well-written and enticing prose. Writing long documents—even technical ones—and choosing the right words is an art form similar to creating a painting or writing a piece of music. Your writing will go better if you find a *secluded* spot where you'll have absolutely no interruptions. The telephone, your e-mail terminal, or other people stopping by will almost certainly break your concentration during writing and interrupt the creative process. Go wherever you can find privacy. Since most of you do not have private offices with lockable doors, I might offer suggestions as to where you can go to write: the company library, cafeteria (during off hours), a bench in your lab, or even a corner of the company lobby. Despite the public nature of these places, they are socially and functionally out of the way and often allow you to avoid interruptions and distractions while writing.

"Now, here are the key steps for writing a long technical document." Fran handed out the following guidelines in hardcopy form, then discussed each item in detail:

A. Plan the Writing Task

1. Gather information about the writing project

"Assemble the results of tests, experiments, design calculations, user specifications, and all other available material. Gather reference citations, figures, and graphics, if applicable. In short, have everything at your disposal before you begin the writing task. And as I mentioned before, you *must* find a secluded place to work."

2. Define the reader

"Decide who will be reading your document. Different readers will have different technical levels and engineering backgrounds, from much more than your own to virtually none at all. It's important to know the technical level of your reader so that you can set the *tone* of the document. If you were reporting on loading tests for the suspension system for the electric car, for example, and the reader was one of the non-technical corporate investors, you would not include a lot of material on spring constants, test methods, or Young's modulus of elasticity. If the report were for engineers inside the company, you might very well want to include these items.

"Regardless of the technical level, you also must keep in mind how much the reader is likely to know about the topic. If the reader knows little, you may want to include more background detail. If the reader is fluent in the topic, you may want to omit needless detail. Also be aware of what the reader will do with the document. Will it be redistributed? Will someone else also read it? Answering these questions will again help set the tone of the document."

3. Make notes

"This next trick is one I'll share with you. It's been used by a lot of the great writers who always seem to get their documents to read just right. On a piece of lined paper, make random, 'stream-of-consciousness' notes—one-line reminders—of *anything* that might need to go into the document. Include the obvious essentials as well as the possible needless trivia. Many writers find it more effective to perform this step with good old pencil and paper rather than on a computer because the act of keyboard typing is known to occupy a sizable fraction of brain activity, leaving less for creative and organizational activities. Regardless of which method you choose for recording your ideas, however, the key is to not worry about the order in which you write things. Just dump them out for further scrutiny at a later step in the writing process.

4. Topic headings

"Your next step should be the formation of the overall structure of the document. To accomplish this task, you should write down the topic headings that you know will need to go into the finished work. Again, you should write these items down in random order, paying no attention to how they will be structured. When you're done with your list, examine each topic heading to see if additional headings come to mind. Delete irrelevant headings, group remaining headings into the main topic areas of the document, and arrange the topics in a suitable order. It's at this point that the main structural framework begins to take shape. When ordering the topics and deciding upon the composition of the paper, consider which order of presentation is the most interesting, logical, and easiest to understand.

"Next, my favorite part: take a break. Often this step means getting something to eat or going out for coffee. It's important to clear your mind before beginning the actual writing process."

B. Write the first draft.

"If you've done your homework in Part A, you're now ready to begin the actual writing process. As I've already emphasized, it's important that you find a quiet place that will permit continuity of thought. Make certain that you will not be interrupted.

This factor is one of the most important to good writing and one that becomes increasingly more difficult as your career advances.

"Start to write. Write anything at first—the important thing is to get words down on paper. Use a word processor with a full-screen editor or write by hand and type later, whichever method suits you. As I've already mentioned, despite the modern convenience of computers, using a keyboard detracts from the creative energy of writing for some people, so don't feel you are chained to your word processor. If you can learn to use it for composing, though, by all means do so. It will save a lot of time otherwise spent in transcribing from handwritten pages.

"At this point, don't be too concerned about spelling or exact phrasing. These aspects of the document can be corrected and modified later, in the revision phase. These days, just about every word processor includes a spelling checker, and you should get in the habit of *always* using it before sending out a document of any kind.

"If the work is short, write the first draft in one work session. If the document is long, divide the work into medium-length work sessions. At the end of each work session, rapidly scan the draft and make only *obvious* changes. Do not do major revisions at this time. When the draft is finished, take a break again to clear your mind.

C. Read the draft

"After your break, when you are no longer intently focused on the document, reread it as if you are seeing it for the first time. Especially check for clarity. Are there vague, confusing, or ambiguous passages? Check for correct tone, for example technical versus non-technical. Is the writing style suitable for both the subject matter and the reader?

D. Revise the draft

"After the initial writing, revise words, sentences, paragraphs, and sections to further refine and clarify the meaning. Never assume that a document is ready for production after the initial writing. As you revise, devote one reading to word usage and grammar only. Set aside concerns about content, and mercilessly slash unnecessary words and sentences. Weigh each word and phrase and keep only those which carry important meaning. Technical writing should be direct and to the point. Is *each* paragraph relevant? Replace complicated phrases with simple words, and limit superlatives. Remove unnecessary words, 'fat,' and details having low information content. During another pass at the document, you should check any factual statements, numbers, and calculations for accuracy. Also be careful to proofread material cited from other documents.

"After you've made your first few revisions, revise, revise, and revise again. Most good writers devote four or more, and sometimes dozens, of rewrites to get a piece to read just right."

E. Review the final draft by asking these questions.

"Put the document aside and come back to it, preferably on another day so that you will not be prejudiced in your evaluation by the intensity of the writing process.

When you next read the document, ask yourself the question, 'Would *I* want to receive what I have written? Will it produce the intended reaction or response from the reader?' If the answer is 'yes,' your document is ready for the outside world."

After discussing the writing of long documents, Fran concluded her workshop with examples of some common writing mistakes made by engineers and technical document writers. "As you know, it's my job as a technical writer to take raw copy from the engineers in the company and turn it into finished product. I'd like to end this workshop by sharing with you a list of common writing errors made by engineers from whom I've received documents for editing over the years. The list is in no particular order, but I hope you will find it interesting and useful when you approach future writing tasks." Fran put up the following slide:

Common Writing Errors
1. The word "this" always needs an object of reference.
Correct: "This problem will be solved by designing a ...."
Incorrect: "This will be solved by designing a ..." (i.e., this ... what?)
2. Sentences including multiple items or ideas should follow parallel construction.
Correct: "Our module will provide data communication, consume minimal power, and satisfy the customer's needs. (All three sentence fragments begin with a verb.)
Incorrect: "Our module will provide data communication, minimal power will be consumed by it, and it will satisfy the customer's needs. (The three sentence fragments don't match.)
3. Use a comma to separate the second part of a sentence only when the second half could stand on its own as a complete sentence.
Do use a comma: "We will supply five commands to the robot, and we will power the robot with batteries." (The second half of the sentence, "We will power the robot with batteries," is a complete sentence.)
Don't use a comma: "We will supply five commands to the robot which will be powered by batteries." (The second half of the sentence, "which will be powered by batteries," could not stand on its own as a separate sentence.)
4. Maintain consistent tense (past, present, or future) as your writing progresses, at least within a given paragraph.
Correct: "The routes will be difficult to change once they have been programmed into memory. This drawback will also apply to future versions of the robot." (future, future)
Incorrect: "The routes will be difficult to change once they have been programmed into memory. This drawback also applies to future versions of the robot." (future, present)
5. "Input" and "output" are nouns ("the input, the output ..."). They are *not* verbs ("the signal was inputted, the data was outputted ...")

EXERCISES

1. The following proposal was written by a student in a design class and outlines the approach to be taken in the design of a system for paging contest participants over a three-minute time interval. It is an example of very poor writing. Rewrite the proposal, taking into account the writing principles and suggestions outlined in Fran's writing workshop.

   The three-minute pager receiver will be based on a simple RC bandpass filter that is tuned to a distinct RF band for each receiver. Additionally, each receiver will tune into a general public announcement band which will broadcast voice messages or tones. The cost will be held very small by constructing our own receiver circuits.

   Power consumption is minimized by sleep mode. In sleep mode, the receiver's PA band amplifier will be disconnected from its power source via a relay or power monitor switch. Detection of a wakeup signal on the wakeup band will close the circuit between the PA band's amplifier and the speaker.

   In addition, preliminary cost research shows that a three-minute countdown circuit and LCD screen can be constructed for under 9 dollars in quantities of 100. A speaker and blinking LED can also be provided at minimal cost.

   The countdown itself would also be initiated by reception of the wakeup signal. The end of the internal countdown would power down the PA band amplifier, or a second detection on the wakeup band would toggle the power off.

   The unit itself could be wearable and styled after a pager or smartcard. We have scheduled a meeting for Tuesday, January 21 at 11 am.

2. The following memo was written by an engineer responsible for designing a parts counting device. The writing style is very poor. Rewrite the memo using the guidelines discussed in Fran's writing workshop.

   During our first conversation with the customer, we came to an initial design for the project and have scheduled a meeting with the clients on Tuesday. For the design, first thing come across is a detector to physically count the parts falling through the sorting mechanism and we generally prefer to use a photosensor. For the counting mechanism, two methods are proposed and yet remain undecided. One of them is to program a PLA whose reprogramming process could be too complicated for the end user. However, its advantage is that the design would be simple and cheap. Another approach is to use a microprocessor to do the counting. However, since our team don't have any experience on this subject before, we are still seeking advice and reference. Finally, when the designated no. of parts are counted, the counter will activate a visual and audio signal which prompts the user that parts are ready for packaging. Then the user can put a plastic bag underneath the container and push a button which opens up the bottom of the container.

   The above would conclude our initial idea and we will come up with more details and specifications after the meeting.

3. Write a memo to your fictitious boss asking permission to attend a technical conference.

4. Write a short technical manual that explains how to operate your VCR.

5. Write a proposal to your student governing body asking for money to start an on-campus amateur radio club.

6. Write a memo to all students in your laboratory addressing the importance of safety procedures and protocol.

7. Write a memo that summarizes the following information relating to data communication protocol:

*DCE = Data Communication Equipment (female connector)
Computer, processor, host: receives data, decodes, establishes communications
*DTE = Data Terminal Equipment (male connector)
Terminal, printer, data board - Sends data and displays output
Parallel Data: 8 or 16 bits of data sent simultaneously with a DR (Data Ready) strobe from the DTE
    to the DCE and a CTS (Clear to Send) signal from the DCE from the DCE
Serial Data: 1 start bit; 8 data bits; 2 stop bits, 14,000 baud (bits audio); no parity
Synchronous data - a clock line must be established between DCE and DTE
Asynchronous data - relies on nearly precise timing and start/stop bits
RS-232 standard (receive-send asynchronous data)
Positive and negative voltages (MARK = 1 = NEG; ZERO = 0 = POS)
Held in MARK state when not in use
DB-25 Connector: pin 1 - shield
                     pin 2 - transmit data to DCE
                     pin 3 - receive data at DTE
                     pin 5 - clear to send (CTS) from DCE to DTE
                     pin 7 - signal ground
Note: The DTE sends data on pin 2

8. The following writing excerpt was handed in by a student as the final report on a laboratory design project. It is an example of poor technical writing. Using the principles and guidelines outlined in this chapter, rewrite the report in the form of a well-written summary memo.

A switch is a device that makes a circuit go from an open circuit to a closed circuit. Such devices are used in houses for light fixtures, to turn the lights off and on. In this design project, two switches are used to build a four-bit binary decoder. In our circuit we want a digital voltage to be one of two values, 0 or 1. These values are binary values, base two values. At the binary value of 1 the switch is up and there is voltage passing through. At the binary value of 0 the switch is down and there is no voltage passing through. In order to see the voltage passing through we put LED's in the path of the voltage from the switch. An LED, light emitting diode, is a round bulb with a flat side that indicates negative and the other side indicates positive. When the LED is put in the path of the voltage it lights up. This occurs when the switch is in the up position and allows the voltage to pass through.

The purpose of our design is to use two mechanical switches for inputs. We made a circuit that lights one of the four LED's for four different combinations of the binary values. We have LED #1 with the binary value of 00, #2 with the value of 01, #3 with the value of 10, and #4 with 11. In constructing our circuit we started by putting the two switches in the middle of the switchboard. Then we put the four LED's spaced throughout the switchboard in a line. Wires were then connected from the switches to the LED's accordingly. Finally two wires were added, one to the positive terminal and the other to the negative terminal, with two leads attached to them from the power supply. This results in each of the four LED's lighting up at the appropriate switch combination.

9. The following memo was handed in by the employee of a small company specializing in adaptive aides for physically challenged individuals. The memo is not written particularly well. Rewrite the memo using the principles and guidelines outlined in this chapter.

To: Xebec Management
From: H. Chew

This project is to work with a 47-year-old individual who has no speech capability and limited physical abilities. The subject groans and grunts to indicate discomfort, displeasure, requesting, and refusing. During dinner, our customer would like us to provide the subject with a means to indicate "I want more," "I want something else," and "I want a drink," etc.

To solve this problem, we had called the customer for more information, and she would give us a video tape that is talking about the older person. We also had a team meeting to discuss the project. At the end of the meeting, we considered that we would design a box which consists of an interface panel and a data control unit. The interface panel would consist of four 2.5" buttons. Each button represents a prerecorded phrase. The sets of outputs from the buttons will correspond to the mode selected. The data control unit consists of the power supply, speech memory, speech synthesizer, and audio amplifier.

10. The following entries were collected by a design team working on a major software project. These notes are to be used by the team to write a summary report to the project manager indicating the features that the finished product must have. The software is to be a voice synthesizer system that will enable individuals with impaired speech capability and limited motor skills to communicate by way of a simple computer mouse. Using the rough notes provided by the design team, write the finished report to the project manager.

Topics covered include: the Alarm, Requests, and Greetings portions of the user interface.

It should be noted that the Alarm portion is, potentially, the most important of these topics. The product should communicate feelings of alarm swiftly and concisely.

The Requests frame should be configured to allow the user to express his most common requests with little effort.

The Greetings frame must have the most common greetings and be designed in such a way as to make these easily accessible.

All the frames should give the user the option to configure them any way he wishes. This certainly includes changing the button labels. It might also include changing the order and/or location of the buttons. Configurable options will be a team decision.

The Alarm frame will be made up of perhaps five buttons each with a clearly marked cause for alarm. These are considered to be emergency situations and, as such, demand immediate attention. Therefore, the user will not have to select the "Speak" button to voice an Alarm message. Each button will be directly tied to the voice synthesizer. Also, selecting an Alarm message will automatically write it to the message frame. The Alarm messages so far are: Help, Pain, Fire, Police, and Ambulance. If, for example, Help is selected, the message "I need help" is written to the message frame and is voiced by the synthesizer. Similar phrases are generated by selecting the other buttons.

The Requests frame is broken down into categories of buttons. Each category will denote a part of speech such as verb, noun, etc. This is especially useful in sentences that express a request. For example, if the user wished to express the phrase "I want a drink," he would go to the first row (or column) of buttons and select the button marked "I." Other choices are "he," "she," "it," "we," "they," and so on. The next category would contain words like "want," and "need." After selecting one of these, the user would have a choice. One category will give him the option of selecting objects such as "book," and "pen," while another category would have choices such as "outside," or "bathroom." What I'm proposing here is breaking up a sentence such as "I want to go outside" into three parts: "I," "want," and "outside."

The Greetings frame will have four categories of buttons. The first category will have standard phrases such as "good morning," "good evening," "good afternoon," and "good night." The second category will contain "Hello" and "Goodbye." These greetings are so common they deserve their own clearly defined category. The third category will be made up of less general greetings that are more personal to the user such as "how are you" or "how are you feeling." In addition to these buttons (that can be customized by the user), there should be a combo box with additional phrases. This will allow him to be flexible and occasionally use other stored phrases that might not fit in the user interface because of the 640 $\times$ 480 resolution. The fourth box should store the replies to personal greetings. If asked "how are you," the user should be able to say "fine, just fine" if he wants. This category should work in a similar fashion to the personal greetings category. There should be some buttons with common replies and then a combo box with other replies. A nice feature would allow the user to easily move a combo box option to any of the buttons. In other words, the combo box should act like a repository of all the phrases while the buttons are hot keys to specific, commonly used phrases.

An important concept here is one of interactive buttons. It is important that we give the user the ability to customize the user interface. While we may decide to make some buttons unchangeable, the user should have the ability to rewrite the text on most buttons to better fit his needs.

Index